Honey Bee Pests and Diseases

Robert Owen has been a beekeeper for over 20 years. He completed his PhD on *Varroa* in 2022. Robert has advised beekeepers in Nigeria, Benin, Lebanon, Nepal and Bangladesh for USAID. He authored the highly regarded *Australian Beekeeping Manual*, also by Exisle Publishing, and regularly contributes to specialist beekeeping magazines. Robert has extensive first-hand knowledge of honey bee pathogens from many countries.

Jean-Pierre Scheerlinck is an Honorary Professor in Animal Biotechnology at the University of Melbourne. Jean-Pierre is passionate about biosecurity, bee genetics, and invading bee pests. An avid beekeeper, he established the University of Melbourne bee club and has developed and taught beekeeping courses for both beginners and advanced beekeepers.

Mark Stevenson is professor of veterinary epidemiology at the University of Melbourne with expertise in the area of infectious disease epidemiology, spatial epidemiology, and simulation modelling of infectious disease spread. Mark deals with infectious and non-infectious disease threats in a range of species including companion animals, cattle, sheep and poultry, horses and honey bees. He has a particular interest in how lessons learned from controlling disease in one species might be used to improve disease control efforts for others.

Honey Bee Pests and Diseases

A complete guide
to prevention
and management

Robert Owen
Jean-Pierre Y. Scheerlinck
Mark Stevenson

EXISLE
PUBLISHING

First published 2023

Exisle Publishing Pty Ltd
PO Box 864, Chatswood, NSW 2057, Australia
226 High Street, Dunedin, 9016, New Zealand
www.exislepublishing.com

A CiP record for this book is available from the
National Library of Australia.

ISBN 978-1-922539-60-1

Designed by Mark Thacker
Typeset in Minion Pro 10.75 on 15pt
Printed in China

This book uses paper sourced under ISO 14001
guidelines from well-managed forests and other
controlled sources.

10 9 8 7 6 5 4 3 2 1

Disclaimer
While this book is intended as a general information
resource and all care has been taken in compiling the
contents, neither the author nor the publisher and
their distributors can be held responsible for any loss,
claim or action that may arise from reliance on the
information contained in this book.

Contents

Introduction

Fifty years ago, there were few known pests and diseases of the European honey bee, *Apis mellifera*. Today, with bees and bee products being shipped internationally on a large scale, many pathogens that were previously geographically confined are now widespread in both developed and developing countries (Figure 0.1). Some of these pests and diseases are new to *A. mellifera* and, consequently, these bees do not display resistance to them. For example, the *Varroa* mite, which has spread from the Asian honey bee (*A. cerana*) to *A. mellifera*, is causing massive colony losses globally (see Chapter 3).

From Figure 0.1, we see that beekeepers are faced with a bewildering array of pathogens to diagnose and manage. Figure 0.2 shows that it is not only pathogens that cause colony ill health but a range of environmental

Figure 0.1: **Some of the many pathogens that infect bees today.**

Bacteria
American foulbrood
European foulbrood
Spiroplasma

Fungi
Nosema apis
Nosema ceranae
Chalkbrood
Stonebrood

Pests
Varroa
Tropilaelaps
Small hive beetle

Viruses
Deformed wing virus
Chronic bee paralysis virus
Acute bee paralysis virus
Slow bee paralysis virus
Israeli acute paralysis virus
Sacbrood virus
Kashmir bee virus
Cloudy wing virus
Black queen cell
Kakugo virus
Lake Sinai 1 & 2 virus
...

Over twenty known honey bee viruses

Figure 0.2: **Factors affecting honey bee health.**

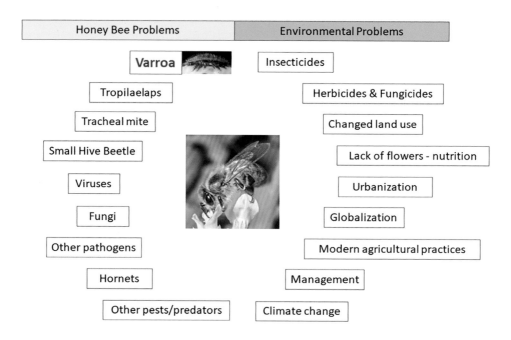

issues. As a consequence, beekeepers today need up-to-date, concise, accurate information on the factors that affect colony health. Previously, much of this information has been locked away in difficult-to-read academic books and journals. This book synthesizes, updates and clarifies the knowledge currently available on bee pests and diseases. It provides, in an interesting and readable form, the information needed by beekeepers to keep their colonies healthy and strong.

Key to the effective management of pathogens is the use of well-proven integrated pest management (IPM) techniques. This book emphasizes sound IPM techniques in every chapter. A firm understanding of integrated pest management, and its use by every beekeeper, will lead to healthy colonies and a healthier future for our bees.

Summary of chapters

The book covers the following topics.

Chapter 1: Integrated pest management (IPM)

Many current beekeeping practices, particularly those related to managing *Varroa*, are harmful to bees, and their continued use is unsustainable. To a greater or lesser extent, most beekeepers practise some form of IPM; this chapter expands on the range of IPM techniques that beekeepers can use to manage the health of their colonies.

Chapter 2: How pathogens infect bees

Most beekeepers understand that their colonies may become infected or infested with pathogens. Often, these same beekeepers do not know how colonies become infected or how pathogens are transmitted. This chapter summarizes how different pathogens can infect a colony and cause illness.

Chapter 3: *Varroa*

Varroa destructor is today's principal threat facing the European honey bee, *Apis mellifera*. A significant amount of research has gone into understanding the role of the mite in colony health, as well as its role in the transmission of viruses such as deformed wing virus (DWV), and black queen cell virus (BQCV). Also, much has been learned about managing *Varroa* by beekeepers on the 'front line' of controlling the pathogen in their colonies and by researchers. This chapter provides a detailed explanation of the biology and nature of *Varroa destructor*, along with detailed explanations of how to manage an infestation.

Chapter 4: **Other parasites**

Varroa is not the only mite to infest colonies. *Tropilaelaps* and tracheal mites also parasitize honey bee colonies. If *Tropilaelaps* were to enter North America, Europe, Australia, New Zealand or any other region with a warm climate outside of its natural home in parts of Asia, the consequences for local beekeeping would be much more severe than has been the case with *Varroa*. Although little is heard about tracheal mites today, historically they have negatively affected colony health. No discussion about the role of mites on bees would be complete without mentioning them.

Chapter 5: **Brood diseases**

The most visible signs of colony ill-health are associated with brood diseases. The chapter discusses two serious bacterial diseases, American foulbrood (AFB) and European foulbrood (EFB), and two fungal diseases, chalkbrood and stonebrood. Chalkbrood is a common but manageable illness of colonies, while American foulbrood, although not that common, is considered the next most serious health concern for bees, after *Varroa*.

Even though insects do not have adaptive immunity (i.e. T and B cells), which in general is restricted to vertebrates, there appears to be some form of trans-generational immune priming (TGIP) protecting the offspring when the queen is exposed to pathogens and passing limited immunity to her offspring. This would allow some form of vaccination although the mechanism is very different than in higher organisms with a full immune system. The recent announcement of a vaccine trial for AFB is an exciting step in the management of this and other serious pathogens, including viruses. Beekeepers will watch with interest as the trial progresses.

Chapter 6: **Adult diseases**

The primary pathogens of adult bees are *Nosema apis* and *Nosema ceranae*. These two pathogens are single-celled parasites of the honey bee midgut that can weaken individual bees and entire colonies. *N. ceranae* is a relative newcomer to *A. mellifera* and, according to some researchers, causes significant economic loss to beekeepers.

Chapter 7: **Viruses**

There are now over twenty known viruses of the honey bee. Some, like sacbrood virus (SBV) black queen cell virus (BQCV), and deformed wing virus (DWV), are well known among researchers and beekeepers. Others are less well known, and their effect on bees and colonies is more challenging to quantify. This chapter summarizes what is known about viruses and how they infect bees; it also details some of the more common viral infections.

Chapter 8: **Pests**

Both greater and lesser wax moths are frequent visitors to the hive. Although they do not directly harm bees, they spin silk mats between frames, trapping bees and making movement through the hive difficult, if not impossible. In some countries, such as India, they cause significant economic damage. Another pest is the small hive beetle (SHB), a native of southern Africa; the mite is causing severe economic damage wherever it goes. Another similar species, the large hive beetle (LHB), although not yet found outside its native home in southern Africa, may cause significant economic loss should it spread more widely.

Braula, a wingless fly, which lives attached to a bee's head, is a common pest. *Braula* does not damage or parasitize any stage of the honey bee life cycle and thus does not cause economic loss.

Chapter 9: **Other problems**

Pathogens are not the only problem faced by honey bees. Other causes of loss, stress or ill health include chilled brood, dampness, drone-laying queens, and multiple eggs per cell. These concerns are detailed in this chapter, together with a discussion on Colony Collapse Disorder (CCD).

Chapter 10: **Other types of bees and hornets**

Although most beekeepers keep the European honey bee, *Apis mellifera*, there are several other species of honey bees found globally, mainly in South East Asia. These include the Asian honey bee, *Apis cerana*, a valuable provider of honey and pollination; the giant honey bees *Apis dorsata* and *Apis laboriosa*; and the smaller species. In southern Africa, we find some very aggressive bees, such as *Apis mellifera scutellata,* as well as the relatively docile Cape bee, *Apis mellifera carpensis*, which can cause economic loss for different reasons. In the Americas, Africanized bees are a severe public health concern due to their extreme aggressivity.

Chapter 11: **Epidemiology**

Although many scientists and beekeepers focus on the colony or individual bees when studying health, honey bee epidemiologists look at disease transmission across large geographical areas and consider broader environmental issues, such as climate change, to understand colony health. The understanding of the epidemiology also leads to formulating policies that affect all beekeepers in their respective regions.

Appendix A: **Diagnostic table**

An easy-to-read chart summarizes the primary bee diseases, their identification and management.

Appendix B: *Varroa* **control options by seasonal phase**

A summary of standard *Varroa* management techniques in table form.

Appendix C: **Integrated pest management table**

A table of IPM management practices.

Every effort has been made to ensure this book is correct and up to date. However, if any errors are detected or you have suggestions for improvement, please send them to: BeeDiseases@gmail.com

Robert Owen
Melbourne, Australia

Integrated pest management (IPM)

Figure 1.1: **Checking brood frame for disease.**

Introduction

Studies of the health of honey bees have demonstrated that many of the problems experienced globally result from poor management practices used by beekeepers. A more enlightened approach to keeping bees would result in healthier colonies and a more sustainable future for the industry. The following chapters on the treatment of pests and diseases list ways each infection or infestation can be managed but not necessarily eliminated. Many of these treatments, particularly for pests like small hive beetle and *Varroa*, involve introducing toxic chemicals into the hive. Many beekeepers and scientists believe implementing this solution has caused more harm to bees than the initial infestation. As a result, a sizeable number of beekeepers are now turning away from the large-scale introduction of chemicals and using instead a set of management practices loosely called integrated pest management, IPM. Although most beekeepers, to a greater or lesser extent, use IPM techniques to manage pests and diseases, in most cases, enhancing these techniques will improve the health and honey-producing capacity of the colony.

IPM is not the same as organic farming practices. Unlike organic farming, IPM includes synthetic chemicals for pest management if necessary. IPM, however, aims to minimize the use of chemicals in the hive (Figure 1.2), and there are excellent reasons for this:

- **Chemicals often harm bees, the environment and other animals and plants, and they may persist for years in wax or soil.**
- **Chemical residues may affect the purity of the products sold, possibly rendering them unhealthy or undesirable in the marketplace.**

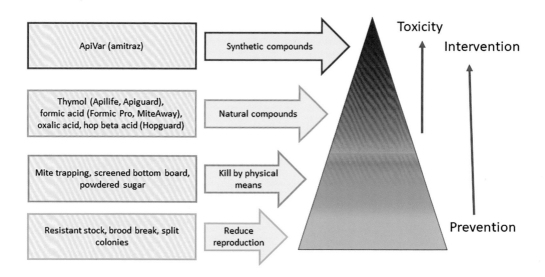

Figure 1.2: **The IPM pyramid for *Varroa* mite control.**

- Chemicals are expensive, and beekeepers can spend much of their income on these substances.
- Chemical resistance by pests or other infectious diseases develops more slowly when bees are exposed to fewer pesticides or antibiotics. Slowing the development of resistance extends the useful lifespan of existing chemical controls.

Consumers are increasingly demanding chemical-free foods. Many consumers are prepared to pay a premium for healthy food; thus, the consumer, the honey bee and the beekeeper all reap the benefit. While IPM is not an organic practice where no chemicals are used, it can nevertheless give the consumer additional confidence that any chemical contamination is at levels below accepted safety thresholds.

Development of integrated pest management practices

Beekeepers in North America, Europe and elsewhere are finding that the inappropriate use of miticides to manage *Varroa* is now causing more problems than the initial infestation. A miticide is an insecticide used to control mites such as *Varroa*, *Tropilaelaps*, or tracheal mites (*Acarapis woodi*). An insecticide is a general term for a chemical used to control all insects, but its use in beekeeping is usually restricted to managing *Varroa* and small hive beetle.

An IPM plan needs to consider all methods that have proven to be successful.

Pests and diseases requiring attention

The eight most important pests and diseases of the honey bee globally are:

Mites
- *Varroa: Varroa destructor* and *Varroa jacobsoni*
- *Tropilaelaps: Tropilaelaps clareae* and *Tropilaelaps mercedesae*
- Tracheal mite, *Acarapis woodi*

Insects other than mites
- Small hive beetle: *Aethina tumida*

Fungi
- *Nosema: Nosema apis* and *Nosema ceranae*
- Chalkbrood: *Ascosphaera apis*

Bacteria
- **American foulbrood:** *Paenibacillus larvae*
- **European foulbrood:** *Melissococcus plutonius*

It is essential to control all of these disorders, and the best way to achieve this is by using an appropriate IPM plan. By keeping the primary diseases listed above under control using IPM, the colony can be left to naturally suppress other lesser diseases, such as the different viruses that infect bees.

Integrated pest management practices

IPM was developed to respond to the overuse of chemicals across all areas of agriculture and animal husbandry. IPM has proven to be a successful pest and disease management technique. The four basic practices of IPM explained below are well understood, and they should become part of the beekeeper's everyday work practice:

Awareness

Each beekeeper, through reading, online searching, talking with other bee-keepers, joining clubs or attending courses, needs to be aware of the pests and diseases that their hives may harbour together with their signs. This awareness should extend even to those hives owned by other beekeepers in the area and to a general understanding of the health of local feral colonies.

Monitoring

Beekeepers need to monitor their hives for the presence of pests and diseases and to evaluate the severity of any infection or infestation.

Treatment thresholds

Each beekeeper must know the acceptable levels for an infestation in their hives before treatment starts. Once you have evaluated the level or severity of infection or infestation, you need to determine if the level is above the threshold to start applying one or more of the possible treatments. It is best to frequently monitor for pests and diseases so that you are not surprised by a high level of illness in an infrequently inspected colony. A record can be kept following inspections that details the pests found, the possible level of infection or infestation, and any action to be taken.

Solutions

The next step is to select the most appropriate treatments or solutions and to decide the optimum level of that solution carefully. The possible range of treatments will usually include applying chemicals to the hive. However, the selected solution may not involve the use of chemicals but, instead, focus on other non-chemical solutions such as:

- **requeening**
- **supplementary feeding**
- **quarantining**
- **reducing stress**
- **SHB traps**
- **vented bottom boards**
- **reducing condensation within the hive**
- **or other proven management techniques.**

An essential part of IPM is to continuously monitor the hive for the presence of pests or diseases. Once you have implemented what you regard as the optimum solution, you need to keep tracking the hive to determine if your selected treatment is working or not. If it is working and the level of infection or infestation is reducing, you can consider the best time to wind down treatment. If the selected treatment does not result in a reduction of

Figure 1.3: **Healthy capped brood.**

the disease or the infestation appears to be worsening, you need to reconsider the treatment chosen and decide what else can be done. Thus, IPM requires constant monitoring and re-evaluation of selected treatments to determine if they are working and what changes need to be made to control the pest or disease.

IPM actions

There is a range of IPM treatments for the major pests and diseases that affect the honey bee. The three broad classifications of management actions are:

Biological control

Biological organisms can be utilized for the control of pests. There are naturally occurring predators, parasites and bacteria that prey on insect pests. Another form of biological control is breeding bee strains resistant to specific diseases; for example, varroa sensitive hygiene (VSH) bees resistant to *Varroa destructor*.

Cultural control

Cultural control occurs when the hive's environment is manipulated to reduce pests and diseases. This may involve weed control, placement of hives, waste removal, traps and general hygiene.

Chemical control

An IPM plan that is well thought through and executed not only aims to slow the development of resistance to a chemical treatment but also considers that some recommended treatments will not be effective. Some chemicals are more appropriate for treating an infection or infestation than others. It is essential that only approved chemicals are used and that their use follows the manufacturer's recommendations. When it is believed that a chemical is needed to back up biological and cultural control, it should be carefully selected and used. The risk here is that if a particular pathogen has become resistant to its recommended chemical treatment, a minority of beekeepers will use unapproved chemicals, often obtained overseas, to treat their hives. Some of these treatments are unapproved because they are toxic to bees and harmful to the broader environment. This scenario will not mean the end of the IPM plan but rather that other essential parts of the program will come to the fore.

How pathogens infect bees

Transmission of pathogens

Honey bees are infected and infested with various pathogens, including fungi, bacteria, viruses, beetles and mites. Bees have lived for millions of years and, over time, have developed a range of strategies to counter infections and infestations. Individual bees, however, possess a lower level of individual resistance to diseases compared to most other insects. They manage with lower personal resistance because, as a colony, they have developed effective strategies to minimize pathogens within the colony.

One of these strategies is the use of propolis, which is antibacterial and acts as a disinfectant. Although many beekeepers go to great lengths to remove propolis from the hive, studies have shown that colonies in hives with more propolis are less likely to suffer from diseases. Another strategy bees employ is to move dead bees away from the colony or for sick bees to leave the colony to die, minimizing the risk to the colony. These strategies are effective, although this chapter will only discuss how pathogens spread to individual bees.

However, other behaviours, such as robbing, increase the likelihood of disease spread.

Beekeeping practices

A problem that has emerged during the last 100 years is that beekeepers have scaled their interventions massively, thus directly and indirectly impacting the colony's life. These large-scale interventions include practices such as migratory beekeeping in which colonies from different apiaries are moved near to each other, exposure to insecticides and fungicides during mass pollination events, and the inappropriate use of miticides to manage *Varroa*. Also, unsustainable agricultural practices, such as the excessive

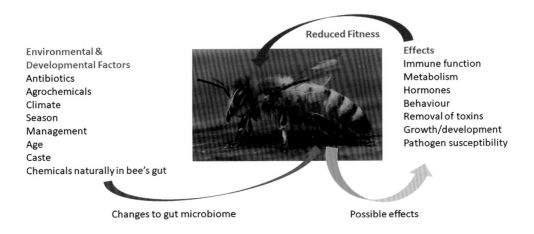

Figure 2.1: **Changes to the bees' microbiome, or gut bacteria, can have serious consequences for the colony.**

use of insecticides, herbicides and monocultures, have harmed bees. A serious effect of agrochemicals on bees is the disruption of 'good' gut bacteria that are critical to the bee's immune system and food digestion (Figure 2.1).

Global trade

The increase in global trade has brought previously unknown pathogens to infect and infest bees in Europe, North America and other parts of the world. These include *Varroa destructor*, *Nosema ceranae*, small hive beetle (*Aethina tumida*) and tracheal mite (*Acarapis woodi*). Another negative aspect of human intervention is that American foulbrood (*Paenibacillus larvae*), is more prevalent in managed colonies than in feral colonies, possibly spread by dirty equipment or the inappropriate use of antibiotics to suppress symptoms but allow the disease to spread. Thus, the European honey bee, *Apis mellifera*, faces a range of infections and infestations that are new to it.

Oral-faecal transmission

The primary route for the transmission of pathogens is oral-faecal (Figure 2.2). Oral-faecal refers to a virus, bacterium or fungus entering a bee through its mouth, perhaps during cleaning or by eating contaminated food. The pathogen subsequently lives and thrives within the bee, resulting in the excretion of live pathogens within faeces. When infected faeces are deposited within a cell or elsewhere in the colony, adult bees remove the faeces, infecting themselves as well (Figure 2.3).

There are several infections whose primary mode of transmission is oral-faecal; the most common are *Nosema apis* and *N. ceranae*. Most bees will leave the colony to defecate away from the hive or nest. During winter or other adverse weather conditions, however, when bees cannot leave the colony, many will defecate within the hive, further spreading the pathogen.

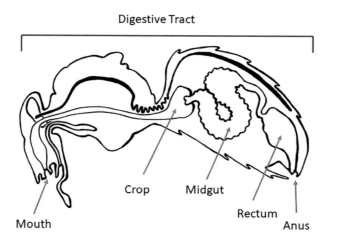

Digestive Tract

Crop   Midgut

Rectum

Mouth   Anus

Figure 2.2: Internal diagram of an adult bee, showing the digestive tract from the mouth to the anus. The mouth is the primary entry point for pathogens. The crop, or honey stomach, is used to store nectar collected from flowers. The midgut is the part of the digestive tract responsible for the digestion of food. Most of a bee's pathogens are here, and the midgut plays an important role in the immune response. The exit point for most pathogens is via the rectum to the anus.

Figure 2.3: **Nurse bees cleaning cells may inadvertently become contaminated by infected faeces.**

Nosema often increases during winter and early spring because of this. Adult bees may have diarrhoea caused by indigestible compounds in honey or pollen or by infection with nosemosis (as infection with either species of *Nosema* is called). These bees may defecate inside the colony, spreading any pathogens they may have. Acute bee paralysis virus (ABPV), chronic bee paralysis virus (CBPV), deformed wing virus (DWV) and Kashmir bee virus (KBV) have also been found in excreta. Black queen cell virus (BQCV) has been detected in the gut of bees, and this may be passed during defecation.

Some pathogens act in concert. BQCV, for example, needs a parallel infection with *Nosema apis* to be able to infect a queen.

Trophallaxis

The process in which an adult bee feeds honey directly to another adult worker or drone is called trophallaxis (Figure 2.4). The honey being fed may have been infected before being fed to the other bee, or pathogens from a previously infected worker may also infect the bee being fed. As an example, trophallaxis may spread the Israeli acute paralysis virus (IAPV).

Nurse bees feed the larvae royal jelly and bee bread, a mixture of royal jelly and pollen. Royal jelly is produced in the hypopharyngeal glands of nurse bees and may carry pathogens from an infected nurse bee (Figure 2.5). Alternatively, the pollen used to make bee bread may have been carrying the pathogen. Both are frequent routes where larvae may become infected. The larval cell wall may also carry the pathogen from the previous brood and so infect the new larvae.

If the hypopharyngeal gland of a nurse bee contains Israeli acute

Figure 2.4: **A bee feeding honey to another bee (a process called trophallaxis) can pass pathogens to the bee being fed.**

Figure 2.5: Larvae fed royal jelly or bee bread may become infected with pathogens.

paralysis virus (IAPV), acute bee paralysis virus (ABPV), deformed wing virus (DWV), or cloudy wing virus (CWV), these viruses may be passed to larvae via royal jelly (Figure 2.6).

Viruses that have been found in honey, pollen and royal jelly include Kashmir bee virus (KBV) and sacbrood virus (SBV). *Apis mellifera* filamentous virus (AmFV) has been detected in honey and pollen, while Israeli acute paralysis virus (IAPV) and Lake Sinai virus (LSV) have been found in pollen. Kashmir bee virus (KBV) has been detected in larval food, possibly originating from nurse bees.

At the larval stage, when the brood is still forming, the digestive tract has not fully developed. It is believed that the development of the digestive tract in the larva is not completed until after all the royal jelly and bee bread have been eaten, so as not to pollute food with faeces. A few days after the larval cell has been capped, the pre-pupa completes the development

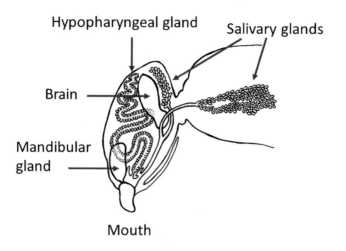

Figure 2.6: **Diagram of a bee's head showing the hypopharyngeal and mandibular glands, which may contain viruses that can be passed to larvae while feeding royal jelly. Also shown are the brain and two salivary glands.**

of its intestinal tract and excretes the stored faeces. If the larva is infected with a pathogen, its excreta may infect any adult bee cleaning the cell. Alternatively, the pathogen may remain on the cell's surface, ready to infect any subsequent larvae that live there.

Forager bees visit flowers to collect nectar and pollen and may leave infected faeces, or pathogens, on the pollen, or they may pick up pathogens left there (Figure 2.7). Along with viruses, bacteria and fungi, *Varroa* mites may also be taken back to the colony, where they can infest a previously uninfested colony or increase mite numbers in an infested colony.

As mentioned, the oral-faecal route is the primary method of disease transmission within a colony. Before they become forager bees, many workers spend part of their lives cleaning cells, comb surfaces and internal surfaces of the hive or nest, ingesting pathogens as they perform this critical activity.

Figure 2.7: **Forager bees may pick up pathogens, including *Varroa*, when visiting flowers.**

Figure 2.8: **A bee's mouth with proboscis extended ready for feeding.**

Grooming

Another cause of infection is grooming. Adult workers frequently groom other bees, including to remove *Varroa*. Pathogens on the surface of one of the bees may pass to the surface of the other bee, where they may be ingested, resulting in infection.

Mating

Virgin queens fly to drone congregation areas to mate with drones from other colonies. Infected sperm may be venereally transmitted to the virgin queen and stored in her spermatheca to fertilize eggs when she returns to

Figure 2.9: **Drones may pass pathogens to virgin queens venereally during mating.**

the colony (Figure 2.9). When the queen lays eggs, there may be pathogens either from the queen or infected sperm. These pathogens need not be transmitted inside the egg but may reside on the surface of the egg. The result will be identical: infected larvae.

Drifting and robbing bees

The drift of both workers and drones to another colony is a frequent occurrence. Drifting bees may take pathogens from their colony to the new colony, which may become infected if it does not already have the pathogen. Robbing bees can take pathogens from the colony being robbed to their home colony, which occurs when a colony has collapsed due to *Varroa* infestations. These colonies may be robbed by bees from other colonies for several weeks after their collapse, resulting in a 'Varroa bomb' when possibly thousands of mites are inadvertently taken by opportunistic robber bees back to their home colonies.

Transmission by *Varroa*

Varroa's infestation of European honey bee colonies dramatically altered the infection landscape (Figure 2.10). Up until this time, the oral-faecal route was the dominant transmission method for the spread of pathogens. Because most pathogens infect bees via the digestive tract, bees have developed an effective system to manage most pathogens in the gut. However, *Varroa* bypasses the oral-faecal route and injects viruses and bacteria directly through the skin of the larva or bee, into the haemolymph (fluid analogous to blood) and/or the fat bodies. Deformed wing virus, Kashmir bee virus and acute bee paralysis virus are known to be transmitted this way. Once inside the haemolymph, the pathogen has evaded most of the bee's immune system and, depending on the virus or bacterium, may flourish, resulting in sickness or death of the bee and, possibly, the colony. Many pathogens, which were previously minor irritants of bees, such as deformed wing virus, have become causes of colony death due to transmission by *Varroa*.

Human activity

Humans are responsible for many of the current disorders afflicting bees. Activities include the reuse of infected equipment and tools, crowding colonies into apiaries and thus enabling pathogens to spread more easily, providing inadequate nutrition so colonies do not obtain essential nutrients, and frequent moving of colonies inadvertently from infected to uninfected areas. In addition, the spread of diseases over vast areas has accelerated due to the transportation of huge numbers of hives to pollinate specific crops. For example, an estimated 1.7 million hives are used for almond pollination in California, requiring the transport of hives across the United States

Figure 2.10: Varroa has dramatically changed the European honey bees' pathogen landscape. (a) Phoretic mite on worker bee. (b) Mature mite on a pupa.

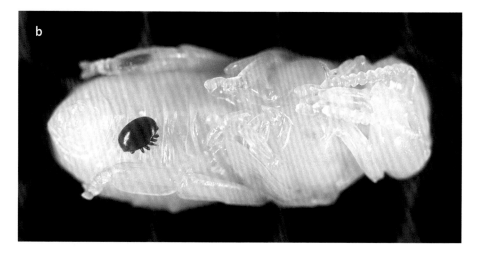

between the east and west coasts every year.

Many people correctly associate the inappropriate application of pesticides with colony ill health. Pesticide poisoning, however, is only part of the story. Many pesticides cause the bee to be more susceptible to pathogens than would generally be the case. Even sub-lethal quantities of pesticides can have long-lasting health consequences for both individual bees and colonies. When different pesticides accumulate over time in beeswax, sub-lethal effects may increase, further harming the colony. Harmful effects of pesticides, even at low concentrations, include an impaired ability to digest food, impaired effectiveness of the bee's immune system, the queen laying fewer eggs, drone sperm being deformed, and poor navigation so that fewer forager bees return to the colony.

Horizontal and vertical transmission of pathogens

Two terms frequently used when discussing the transmission of pathogens are horizontal and vertical transmission. A pathogen is transmitted horizontally when the infection is passed between members of the same generation. An example would be the spread of American foulbrood either within a colony, from an infected worker to a larva, or between hives via contaminated equipment used by a beekeeper. *Varroa* moving to another colony, attached to the body of a robber bee or drifting drone, is also horizontal transmission.

Vertical transmission occurs when a queen passes a pathogen to her offspring via the egg. Deformed wing virus and sacbrood virus may be transmitted vertically from queen to larva.

Scientists believe that pathogens that are transmitted horizontally may lead to the pathogen becoming more virulent over time. An example may be the horizontal transmission of the deformed wing virus by *Varroa*. The main mechanism for the spread of the mite is the collapse of an infested colony and robber bees taking mites back to their own colonies. Thus, the collapse of a host colony is an important part of *Varroa* transmission and reproduction. This may have led to *Varroa* and deformed wing virus becoming more lethal over time.

Pathogens that are transmitted vertically, from queen to egg, rely on infected colonies remaining healthy in the presence of the disease. If a pathogen that was transmitted vertically led to the death of, say, larvae, the colony would collapse, and the pathogen would not survive. Pathogens that are transmitted vertically are likely, over time, to become less virulent so as not to lead to the collapse of a colony.

CHAPTER 3

Varroa

Varroa destructor: introduction

When *Varroa*, tracheal mite (*Acarapis woodi*) and small hive beetle (*Aethina tumida*) were detected in honey bee colonies in the United States and Europe, there was a concerted effort to identify all the mites and pathogens that live with honey bees. To date, a range of diverse mites has been identified, although only three cause economic losses to beekeepers: *Varroa destructor*, tracheal mite and the *Tropilaelaps* mite (*Tropilaelaps* species). There is also the small hive beetle. *V. destructor* is causing severe economic loss to beekeepers across the world. *Tropilaelaps* is mainly restricted to Asia and is not yet a problem to beekeepers in Europe, North America or

Figure 3.1: **Adult female *Varroa* mite on pupa.**

Australasia. However, *Tropilaelaps* is causing *A. mellifera* colony losses in South East Asia.

Two species of *Varroa* infect the European honey bee:

- *Varroa destructor*
- *Varroa jacobsoni*

Varroa destructor is believed to have jumped species from the Asian honey bee, *A. cerana*, to the European honey bee, *A. mellifera*, before the 1950s. By the 1970s, *V. destructor* had migrated to Europe and from there to the United States. More recently, *V. jacobsoni*, a mite often associated with *A. cerana*, has been found to occasionally also infest European honey bees, with this first occurring in Papua New Guinea. This development is causing concern as the new parasitic mite adds additional stress to an already stressed species.

Spread

At the local level, the primary cause of *Varroa* spread is scout or forager bees robbing colonies in search of honey. Other causes include infested bees drifting to other colonies, or foraging bees leaving mites on visited flowers where other foragers pick them up. Requeening with an infested queen, or the splitting or merging of colonies, can also spread the mite. Beekeepers can help control *Varroa* by ensuring that sound management principles are used to minimize the mite's spread. In the United States, migratory beekeeping spread *Varroa* throughout the country in five years.

If a feral colony is infested with *Varroa*, it may live only about two to three years before the colony dies, possibly by infecting the bees with deformed wing (Figure 3.7) and other viruses. On the positive side, even though infestation by *Varroa* is an expensive and time-consuming problem, a beekeeper with colonies infested with the mite often reports increased honey production with fewer hives. The reasons for increased production vary from the ability and management skills of the individual beekeeper to other factors, such as fewer feral colonies competing for nectar and pollen. Beekeepers often reduce hive numbers to deal with the added workload of managing *Varroa*-infested colonies, which is a direct cost of managing the mite.

Signs of infestation

V. destructor is large compared to many other mites and can be seen with the unaided eye. The width of the female mite is larger than her length, about 1.6 mm x 1.1 mm, with a reddish colour to her body. Adult females move into brood cells to reproduce, and newly emerged female adults are often seen walking rapidly on the surface of brood comb before entering another larval cell to reproduce (Figure 3.3). Individual mites often attach to adult bees, mainly clinging to the bee's abdomen or hidden in the inter-segmental membrane where they feed on fat bodies from inside the bee.

Figure 3.2: **Different stages of male and female *Varroa*. Adult male, top left. Adult female, top right. Growth of immature female mites on the bottom row.**

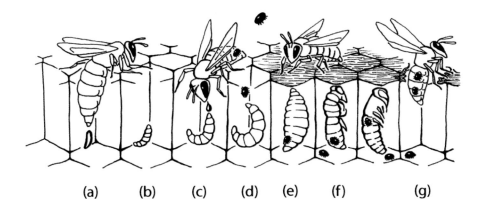

(a) (b) (c) (d) (e) (f) (g)

Figure 3.3: Stages in the reproductive life of *Varroa*. (a) Queen bee lays egg in a brood cell. (b) Shortly after, the bee's egg turns into a larva. (c) Nurse bees feed royal jelly to the larva. (d) Female mite crawls into the cell and waits for the cell to be capped. (e) When the cell is capped, the female mite starts to lay eggs. (f) First, a male egg, then female eggs. The young male mite mates with his sisters. (g) When the emerging adult bee leaves the cell, the newly hatched female mites and the old mother leave with the emerging adult bee, ready to enter a new uncapped cell to repeat the process.

When a mature female mite enters a brood cell, she first lays a male egg and then about six female eggs, approximately one egg every 30 hours, laying about five eggs in worker cells, and six eggs in drone cells. If the initial female mite is the only *Varroa* to have entered the brood cell before it is capped, the male mite she first lays will fertilize his sisters, expanding the number of viable female *Varroa* mites in the colony. If other mature female *Varroa* mites enter the brood cell before it is capped, the male mites can fertilize the daughter mites of other *Varroa* mothers, bringing greater genetic diversity to the reproduction process.

Immature mites are white while adult females are reddish brown; adult males are smaller than females and, as previously noted, are not seen as their entire life cycle is spent inside brood cells (Figure 3.2). Adult male *Varroa* and immature female *Varroa* are not seen outside brood cells as

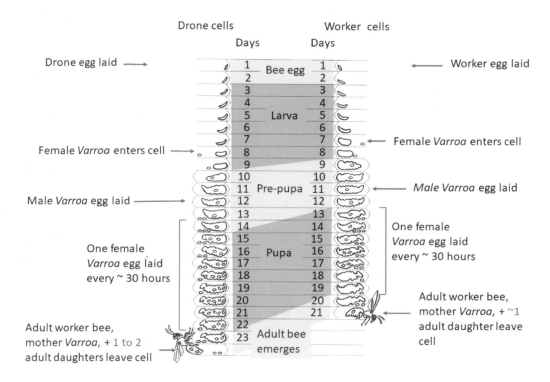

Figure 3.4: *Varroa* has more adult offspring when laying eggs in a drone cell, compared with a worker cell. On left, from top down, reproduction in drone cells: on day 1 queen lays a drone egg. On day 10, drone cell is capped with female mite inside. Day 12, a male *Varroa* egg is laid. Day 14, the mother *Varroa* mite starts to lay female mite eggs. By days 21 and 22, newly hatched female mites are mature and are adult. Day 24, the drone leaves the cell, and female mites leave as well. On right, from the top down: on day 1, a female worker's egg is laid. On day 11, a male *Varroa* egg is laid. On day 13, the mother mite starts to lay female mite eggs. The female adult mite is mature on day 20 and leaves the pupal cell on day 21. The result is that, on average, a *Varroa* mother in a drone cell brings to maturity about two mites, while in a worker cell, usually one or two mites are brought to maturity.

they are unable to survive without the wet and humid conditions found inside the capped cell. The development stages of a *Varroa* mite are:

1. Adult female mite enters an uncapped larval cell, a day or two before it is capped.
2. When the cell has been capped, the adult female mite lays eggs inside the cell.
3. The first egg laid is usually a male; subsequent eggs are female.
4. While inside the cell, each mite egg develops into a larva, and then into an immature male or female adult mite.
5. A *Varroa* protonymph develops, which feeds on the bee pupa. A protonymph is the first stage, after hatching, in the development of many mites.
6. The *Varroa* protonymph moults and emerges as a deutonymph. A deutonymph is the second stage in the life cycle of a *Varroa* mite (Figure 3.5).
7. The deutonymph also feeds on the bee pupa.
8. The deutonymph moults again and emerges as an adult *Varroa* mite.
9. The male mite mates with his younger sisters.
10. The bee pupa emerges from the capped cell.
11. The mother mite emerges, together with her fertilized daughters.
12. Both the mother mite and her daughters attach to adult bees, where they hide and feed.
13. After a few days, the female mites enter an uncapped larval cell, possibly attracted by the smell of the bee larva, and the process is repeated.

On average, a female *Varroa* mite in a capped worker brood cell produces 1.4 to 1.5 mature daughter mites that emerge from the cell. Although the *Varroa* mother lays several eggs, most daughter mites do not reach maturity before the cell is uncapped and die within the recently opened cell. The daughters can remain on nurse bees for up to nine days before entering another brood cell to lay further *Varroa* eggs. The adult honey bee workers emerge from their cells after about twelve days as pupae, while drones emerge after about fifteen days from their pupal cells. Since a fertile *Varroa* mother will lay several eggs inside a capped brood cell, the long gestation period for male drones to emerge as adults means that about 2 to 2.5 mature *Varroa* daughters will emerge from a male drone cell with the drone (Figure 3.4). Also, drone cells are eight to ten times more attractive to *Varroa* looking for a home than worker cells, possibly due to the different odours emitted by drone larvae than worker larvae. Usually, the number

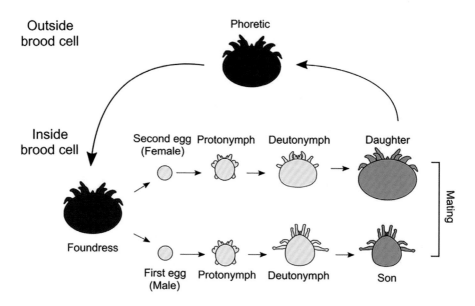

Figure 3.5: **Egg, protonymph and deutonymph stages in the life cycle of** ***Varroa.***

Figure 3.6: (a) A *Varroa* mites on pupa. (b) A mite leaving the cell with a hatching worker bee. (c) A mite on an adult bee (overleaf).

c

of capped drone cells inside a colony determines how many mature *Varroa* adults will emerge and infect the hive.

The mite attacks the young brood in a capped cell by piercing the soft skin of the pupa to feed on its fat body. If only a single mite enters the brood cell, damage to the pupa may not be visible, although the adult bee may have a much shorter lifespan than unparasitized bees. If multiple mites enter the brood cell, the brood pupa may die, and any adult bees that emerge may be deformed with a shortened abdomen or deformed wings. Deformed wings indicate infection with deformed wing virus, but it may be caused by other factors such as chilled brood. In addition, the behaviour of the adult bee may also be affected, such as by poor orientation during flight or difficulty in gathering food. Damage by mites when infesting adult bees is usually restricted to heavy infestations when a substantial proportion of the colony is infested.

Fat bodies in insects perform a similar role to the liver in mammals and are solid organs within a larva or adult bee's abdomen. When a mite is ready to feed, a *Varroa* closes its mouthparts and forces them through the soft tissue of the larvae or adult bee. The closed mouthparts can be thought of as a hard beak, with a straw-like channel through the middle. When the mouthparts are inside the bee, they release a cocktail of digestive enzymes that breaks down the fat body, while still inside the bee, into a slurry they then consume. *Varroa* does this because it is unable to feed on the solid matter so needs to break down, or digest, the fat body while still inside the bee. When the fat body has been digested, the mite can siphon the nutritious slurry, using the straw-like channel inside its mouthparts, back into its body to use as food.

A *Varroa* infestation is more complex parasitism than just mites eating brood fat bodies since the mite also injects viruses such as the Israeli acute paralysis virus (IAPV), or deformed wing virus (DWV) into the brood during feeding. These and other viruses may significantly impact the bee's health, causing paralysis in adults or deformed wings. Infections transmitted by *Varroa* may severely reduce the quality and lifespan of an infected bee, and dramatically speed up the colony's demise. Indeed, virus-free *Varroa* mites need much higher mite loads to kill the colony than virus-infected mites. In Australia, when *Varroa*, free of DWV, entered the country, one apparently healthy infested colony was reported to have 30 mites per 100 bees. The usual lethal number of DWV-infected mites in North America is about 3 mites per 100 bees. However, even in the absence of viruses, mites eating pupal fat bodies may decrease disease resistance since fat bodies are part of a bee's immune system. In addition to storing energy, fat bodies in bees perform a similar role to the liver in mammals and are involved in the processing of many proteins.

Treatment

Where *Varroa* has been established, depending on the mites' viral load, within a few years, the survival of managed European honey bees has only been achieved through intensive beekeeping work. Most colonies of bees have required significant and expensive levels of management to survive, while feral colonies, without treatment, die within two to three years of becoming infested. A tiny number of feral colonies, however, have been found to survive untreated in the United States, Europe and elsewhere. These colonies are still infested with *Varroa*, but the mites are not as harmful to the bees as they once were. In some cases, the mite became less

Figure 3.7: **An adult bee with deformed wings. Deformed wings are often a sign of infection by the deformed wing virus (DWV), transmitted by *Varroa*.**

harmful to the colony; in other cases, the colony became more resistant to infestation.

Since *Varroa* can be spread by the activities of the beekeeper, including the practice of migratory beekeeping, beekeepers need to review how they manage colonies. Appropriate management is one of the key tools to minimize the spread and impact of the mite. Integrated pest management needs to be at the front of every beekeeper's toolbox if *Varroa* is to be managed long term.

Another management technique to control the mite is to disrupt their breeding patterns by inserting drone brood comb, in which the *Varroa* mite prefers to reproduce. When many of the drone cells have been capped, the drone brood comb is removed by the beekeeper, and the capped cells are destroyed, killing the mites before the drones and *Varroa* hatch out.

Many chemicals, such as Amitraz, may be used as both a miticide and an insecticide. The chemicals used to treat *Varroa* vary in their toxicity to both bees and humans. The naturally occurring organic chemicals oxalic acid and formic acid are relatively safe, while others must be applied carefully. Miticides such as Apiguard, Apistan and Apivar should not be used in hives if supers contain honey. The honey may become contaminated and unfit for human consumption.

Miticides, however, are still the most popular way to control *Varroa*. Although using chemicals risks contaminating honey and wax with lethal pesticides, these risks can be minimized if used correctly by following IPM practices and the instructions on the packet. Medicating the hive should not be done unless it is essential. Applying chemicals to a hive as a protective measure without any signs of *Varroa* is counterproductive since toxic residues will be left in the hive.

Overuse of miticides to control mites will cause mites to become resistant to the miticide over a short period. This has already happened with fluvalinate and coumaphos, and in some cases with Amitraz. By rotating three different miticides, the development of resistance can be delayed. See Chapter 1 on integrated pest management for more details.

At the time of writing *Varroa* has not been detected in Newfoundland, Canada or several Pacific Island countries. Judging by overseas experience, once *Varroa* reaches a previously uninfested area and gains a small foothold, it cannot be eradicated, and the spread of the mite to the rest of the region will be rapid.

The long-term treatment of *Varroa* with chemicals, as with all honey bee mites and pests, is not acceptable due to the harm these toxic miticides are causing bees and the likelihood that resistance to the chemicals will appear in the mite population. One long-term solution is to breed bees that are resistant to these mites as well as to other diseases. In the United States, several strains of *Varroa*-tolerant bees are available, which display resistance to *Varroa*. *Varroa*-tolerant bees include the varroa sensitive hygiene (VSH) strain developed by the United States Department of Agriculture (USDA), as well as so-called Russian bees. The use of survivor stock is also a promising solution. Survivor stock is the resistant colonies of feral bees left after *Varroa* has wiped out all of the susceptible honey bee colonies.

Varroa-tolerant strains of queens are effective at producing colonies that can manage *Varroa*. The second generation of the queen, if not mated with drones from the same resistant stock, is not as effective and quickly loses its resistance. As effective as *Varroa*-tolerant queens are in controlling *Varroa*, their adoption by North American beekeepers is not widespread, and more needs to be done to encourage their use. While these bees are resistant to *Varroa*, they are not selected for other characteristics considered necessary by beekeepers. *Varroa*-resistant bees often do not display high honey production, low swarming tendency, gentleness, and quick build-up in spring; hence, they are often outperformed by other strains with these characteristics. Selectively breeding *Varroa*-tolerant strains of queens also needs to be performed at the local level by individual beekeepers and groups of local beekeepers, not just by the more prominent commercial queen breeders. Only by breeding queens that are tolerant of local conditions (pests and diseases as well as climate and food sources) and meet the beekeeper's specific needs will the honey bee industry get a grip on the solution to many of the problems that confront it.

See Appendix B for details of recommended miticides.

Varroa monitoring

The best way to ensure the survival and productivity of your bee colonies is to measure your *Varroa* mite load and treat it as necessary. It is impossible to optimally manage an infestation without knowing the number of mites in a colony. The objective of sampling is to determine the percentage of mite infestation, usually expressed as mites per 100 adult bees and called the infestation rate. The infestation rate harmful to honey bee colonies will differ depending on the seasonal phases of the honey bee colony size, geographic location, and the presence or absence of viruses transmitted by the mites.

The effect of *Varroa* on a colony will vary between countries. Threshold infestation rates in North America are likely to differ from those in Europe or elsewhere. Each beekeeper needs to determine the threshold for treatment, in their country or locality, before deciding on a suitable treatment. In the United States, 2–3 per cent infestation rates should be treated promptly. Additionally, mite counts on sticky boards from a 10-frame Langstroth hive exceeding 59 mites in 24 hours should also be treated promptly.

A sugar shake, alcohol wash, sticky board or drone uncapping provides only a snapshot of any infestation. Tests need to be repeated every three to six weeks to better understand mite loads in your colonies.

The following is a list of popular methods to estimate the infestation rate of a colony.

Sugar shake

Sugar shake monitoring is probably the most widely used *Varroa* surveillance tool in use (Figure 3.8).

1. Remove a brood frame from the hive, not a frame of honey, and check the queen is not present.
2. Brush or shake the bees into a bowl or onto a sheet of newspaper, and wait about a minute for the older forager bees to fly off. This step will substantially increase the probability that tested bees are infested.
3. Funnel or scoop about one cup of bees into a jar.
4. Add approximately one heaped tablespoon of powdered (preferably sifted) icing sugar to a jar.
5. Secure the mesh lid onto the jar (Figure 3.8a).
6. Vigorously shake for one minute, to cover the bees in sugar and dislodge the mites from the bees. Sugar-covered bees must be vigorously shaken, to the extent that some may die. Otherwise, mites may not be dislodged, giving a false negative test. Your arm will get tired, so take a rest after 20 seconds.
7. Set the jar down and wait three to five minutes. Rushing

the process increases the risk that an infestation will not be detected.

8. Invert the jar and **vigorously shake** it like a saltshaker over a pan of water. It is better to shake the jar over water than over a clean plate since any mites will float on top and will be easier to identify. The sugar will also dissolve in the water, making mite detection easier.

9. Add another tablespoon of icing sugar to the jar, shake and roll the bees again for 30+ seconds, and repeat steps 2, 3 and 4 to improve the sensitivity of the test.

10. Inspect the pan of water to see if any mites are floating on the surface. There may also be dirt floating, so glasses or a magnifying glass may be needed.

11. Release bees that have been tested into the top of their colony or at the colony entrance.

12. Since we are estimating the number of phoretic mites per 100 adult bees, if we sampled 300 bees and found, say, nine mites, the rate of infestation would be 9 per 300 = 3 mites per 100 bees. In North America, this level of infestation, depending on the season, may require immediate treatment of the colony with a miticide.

During the initial stages of a *Varroa* incursion, mite counts in a hive are likely to be low. In this case, the jar containing sugar-coated bees must be **shaken vigorously for up to a minute** to maximize the probability that any mites will be dislodged and detected. Many descriptions of the sugar shake technique say to roll the jar of bees gently. Gently rolling the jar is unlikely to dislodge any mites. Although beekeepers are reluctant to kill bees during testing, the jar needs to be sufficiently agitated that some bees under test are likely to be killed or damaged. Without vigorous shaking, an accurate estimate of the infestation level in a colony cannot be made.

To increase testing accuracy if you have only recently started monitoring *Varroa* levels, count the number of bees being tested. The number does

Figure 3.8: Sugar shake *Varroa* monitoring. (a) A sugar shake jar. (b) Adding spoon of icing sugar to jar. (c) Shaking worker bees from a brood frame onto a sheet of paper. (d) Using the folded paper, worker bees are poured into the jar. (e) A cup of bees, usually 300 nurse bees. (f) After vigorous shaking, the jar is inverted, and the contents are tipped into a bowl of water. The mesh lid stops bees from falling into the water. (g) After checking the surface of the water for mites, the contents of the jar are poured back into the brood chamber of the hive.

not have to be exactly 300, although the more bees in the sample, the more accurate the test will be. Adjust the number of bees in the formula to calculate mite load as needed.

Alcohol or soap wash

Although most beekeepers know of the alcohol wash detection method, many may not be aware that dishwashing liquid may be used in place of alcohol, with the same accuracy. In the following description of the alcohol wash, the same steps need to be taken if dishwashing liquid is used. In North America, Dawn washing liquid may be used; in Europe and Australia, the Fairy brand is suitable.

1. Place 300 bees into a jar and cover with isopropyl alcohol, methylated spirits or dishwashing liquid (Figure 3.9a).
2. Shake vigorously for one minute to dislodge the mites from the bees. Your arm will get tired, so take a rest after 20 seconds.
3. To ensure consistency between tests, or colonies, shake each sample with the same vigour, and for the same amount of time, for each test.
4. After shaking, empty liquid into a strainer, with the alcohol falling into a container.
5. Add more alcohol to the jar of bees and vigorously shake again. This increases the accuracy of the test.
6. Empty the alcohol into the strainer and let the liquid collect in a container with the previous liquid.
7. It is important that the mesh of the strainer allows any mites to fall through without allowing bees to fall through.
8. Count the number of mites in the container.

The same method to estimate infestation rate is used as described for sugar shake. If 300 bees are tested and 9 mites are dislodged, the rate of infestation is 9 per 300 = 3 mites per 100 bees.

Figure 3.9: Alcohol wash *Varroa* monitoring. (a) Commercially available alcohol wash equipment. (b) Bees are collected and added to the jar in the same way as collected during a sugar shake test. The jar containing alcohol and bees is vigorously shaken. (c) The inner strainer containing bees is removed, leaving alcohol in the jar. (d) The alcohol is inspected for *Varroa* mites.

Sticky boards

Many beekeepers prefer the sticky board monitoring method because it does not require opening a hive and removing nurse bees. A modified base-board is needed to slide in and out a sticky mat on which dead mites will fall (Figure 3.10).

1. If a commercial sticky mat is used, peel off the protective layer from the mat.
2. Slide the mat into the bottom of the hive under a mesh that stops bees from walking on the mat but which allows mites to fall through.
3. Place miticide-impregnated strips between the frames of the brood box according to the manufacturer's instructions.
4. Leave the sticky mat in place for 24 hours.
5. Remove the mat and count the number of mites.

Depending on your location, the number of mites that fall over 24 hours will trigger treatment. In parts of North America, if 30 mites fall in 24 hours, treatment with miticide will be necessary. If the miticide strip is left longer than 24 hours, a correction to the mite count needs to be made to standardize the count over 24 hours; otherwise, too high a mite count may result in excessive and inappropriate management actions.

Commercially made sticky boards are expensive. Many beekeepers make their own sticky boards by cutting a sheet of Coreflute, cardboard, or other board to fit inside a hive, and covering it with petroleum jelly. Petroleum jelly can be applied using a paint roller.

Drone brood uncapping

If a colony has a large number of capped brood cells, up to 85 per cent of mites may be inside a capped cell, making them undetectable using sugar shake or alcohol wash. A practical method to estimate mites in a colony with a lot of capped brood, say during the early spring, is to remove a frame

Figure 3.10: (a) *Varroa* and debris on sticky board left for 24 hours.
(b) Mesh base for use with sticky board (left), standard solid base (right).

with lots of the larger capped drone brood cells on it. Drone cells are preferred because *Varroa* preferentially moves into drone cells to breed. Also, the long gestation period for a drone allows the mother *Varroa* to lay more eggs in the cell, making counting easier.

1. Remove a frame near the hive's centre that contains a large portion of the drone brood. If the queen is present, place her back in the hive. Shake the remaining bees back into the hive.
2. Using a scratcher, push the fork's prongs through the ends of the capped drone brood cells, and remove as many pupae as possible (Figure 3.11). Note that this will kill the pupae.
3. If possible, repeat the process on about three brood frames. Depending on the size of your apiary, randomly select frames from about three to five hives.
4. Uncap about 100 drone brood cells.
5. Check each pupa for a reddish-brown mite. These can be seen against the white body of a young pupa.
6. Mites are easier to see on young pupae, rather than on older pupae that have taken on the coloration of an adult bee.
7. When pupae have been removed from cells, tap the frame over a white board. Mites that were not attached to pupae may fall onto the board.
8. Also check the bottom of each drone cell to see if any mites are left.
9. Return each frame to the hive.

Figure 3.11: (a) Removing drone pupae from drone comb frame. (b) Killing drone pupae infested with *Varroa*. (c) A scratcher is used to remove drone pupae from capped drone cells.

CHAPTER 4

Other parasites

Tropilaelaps mites – the Asian mite

Introduction

Tropilaelaps mites are similar to *Varroa* and are parasites of honey bees (Figure 4.1). The giant Asian honey bee (*Apis dorsata*) is their natural host, and four species have been documented (*Tropilaelaps clareae*, *T. koenigerum*, *T. thaii* and *T. mercedesae*). *Tropilaelaps* has also been found in colonies of other Asian honey bees, including *A. cerana* and *A. florea*. Two of these, *T. clareae* and *T. mercedesae*, infect the European honey bee, *A. mellifera*.

Varroa has received enormous attention over the past three decades and, as a result, the importance of other parasitic mite species on bees has been largely ignored. Also, because *Tropilaelaps* mites are mainly found in Asia, their threat has been largely overlooked in other parts of the world. If *Tropilaelaps* were to enter Australia, the United States or Europe, the consequences of this incursion would be far greater than that of *Varroa* due to its much shorter reproductive cycle. *T. clareae*, which infests the European honey bee, has a wide distribution in Asia, from Iran to Papua New Guinea. Other species of *Tropilaelaps* have a more restricted geographic distribution.

Adult *Tropilaelaps* mites are light reddish brown and, depending on the species, between 0.7 and 1.0 mm long and approximately 0.6 mm broad. Like *Varroa*, they must enter late-stage bee larvae cells before capping to reproduce. Forty-eight hours after cell-capping, the mite starts laying three or four eggs, each of which will hatch within 12 hours. Five days after hatching within a capped cell, the mite larvae will develop before becoming adult mites.

The developing mite feeds on brood, depriving the larva of the nutrients

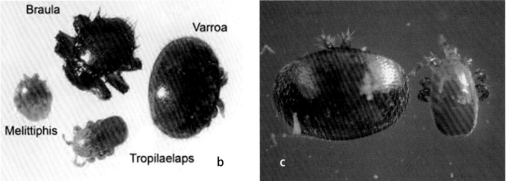

Figure 4.1: (a) *Tropilaelaps* on European honey bee larvae and pupa.
(b) *Tropilaelaps*, *Varroa*, *Braula* and *Melittiphis*. (c) Comparison of *Varroa*
and *Tropilaelaps*.

required for growth. At the same time as the young adult bee emerges from its cell, the new generation of female adult mites (and their mother) leaves the cell in search of fresh hosts to live off. Adult female mites that have left a pupal cell need to find a new larva-containing, uncapped cell within two to three days, or the mite will die.

Tropilaelaps has been shown to harbour both deformed wing virus and black queen cell virus, serious problems for *A. mellifera*.

Spread

The spread of *Tropilaelaps* is similar to that of other mites such as *Varroa* or the insect pest small hive beetle. *Tropilaelaps* mites move quickly and can readily transfer from one bee to another within the colony. To move between colonies, they depend upon adult bees for transport through the natural processes of drifting, robbing and swarming. *Tropilaelaps* can spread quickly through an apiary by poor beekeeping management practices, such as moving infested combs and bees. The rapid diffusion of the mite to new areas occurs when beekeepers transport colonies in search of food or to pollinate crops. Over the last 50 years, *Tropilaelaps* has spread beyond its ancestral area. Indeed, *Tropilaelaps* can flourish in any warm environment where honey bees produce brood throughout the year. *Tropilaelaps* have been reported in Papua New Guinea and Kenya, far from their native home.

The life cycle of *Tropilaelaps* is similar to that of *V. destructor*, but *Tropilaelaps* develops much faster, and the intervals between successive reproductive cycles are shorter. As a result, if both types of mites are present in the same colony, *Tropilaelaps* numbers can build up at a much faster rate than those of *Varroa*.

Signs of infestation

Tropilaelaps is often mistaken for *Varroa*, although distinguishing between the two is straightforward:

- A *Varroa* mite is larger, crab-shaped, and broader than it is long.
- *Tropilaelaps* is about one third as wide as a *Varroa* mite, or 1 mm long and 0.6 mm wide.
- The *Tropilaelaps* mite's body is elongated.
- Adult *Tropilaelaps* mites run rapidly over infested brood combs. A video taken by USDA researchers shows the mites running around on the comb much faster than *Varroa*.
- Unlike *Varroa*, *Tropilaelaps* mites hide in brood cells rather than on adult bees.

Infection by *Tropilaelaps* causes abnormal brood development and the death of both brood and adult bees, causing colony strength to decline, resulting in absconding or even the collapse of the colony. Adult bees damaged during development have reduced lifespans, lower body weight, and wing and leg deformities. With severe infestations, colonies have a noticeable smell of decaying bee remains.

European honey bees (*A. mellifera*) lack the well-developed defences to *Tropilaelaps* that the giant Asian bee (*A. dorsata*) has developed over possibly thousands of years and is thus more susceptible to the effects of the infestation. Also, unlike *A. mellifera*, *A. dorsata* bees can bite and injure *Tropilaelaps* mites and are better at grooming and removing the mites.

Perhaps the most alarming characteristic of the mite is its reproduction rate and breeding cycle. Most *Tropilaelaps* mites live and reproduce inside brood cells, living only about two to three days on an adult honey bee outside of a pupal cell. One to four adult female mites are generally found in a cell reproducing at one time; however, there are reports of mites being found in quantities over three times that many in just one cell. They can quickly outnumber any *Varroa* mites in colonies, hatching in only twelve hours and reaching full maturity in six days. Within 24 hours of leaving a pupal cell, the mites enter other cells and begin reproduction.

Treatment

Treatment for *Tropilaelaps* is similar to that for *Varroa*, although, since this mite has mainly been found in Asia, no treatment has been agreed upon in other parts of the world.

An essential difference between *Varroa* and *Tropilaelaps* is that a mature *Varroa* mite can survive for many months outside of a brood cell by attaching herself to a bee's abdomen and eating its fat body after biting into its abdomen. *Tropilaelaps* mites cannot survive this way since they do not have sufficiently strong mouthparts to bite into an adult bee's abdomen. Thus, one way to control a *Tropilaelaps* infestation is to reduce the brood for several weeks by:

- **removing the queen from the hive for three weeks so that no more eggs are being laid, or**
- **placing her inside a cage where she cannot lay eggs for three weeks while still kept inside the hive.**

Tracheal mite

Introduction

The tracheal mite (*Acarapis woodi*) is a tiny parasitic mite that infests the airways or tracheae of the honey bee (Figure 4.2). They were first identified during the Isle of Wight bee disease outbreak in the United Kingdom in 1919 and were initially thought to cause this large loss of colonies. Tracheal mites were first detected in the United States, in Florida, in 1984.

Infestation by tracheal mite is commonly called acarine disease. Because tracheal mite disease occurred across Europe simultaneously, killing many thousands of colonies, it indicated that it was a new disease in Europe, although its origin is still unknown.

Although many beekeepers believe tracheal mites cause significant economic damage, others believe harmful effects are less severe than initially thought.

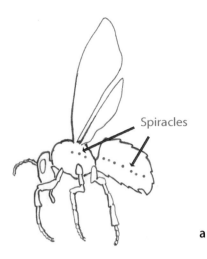
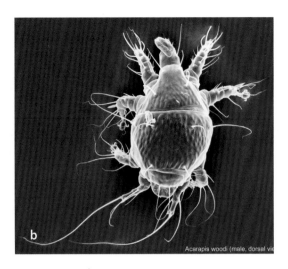

a

b

Figure 4.2: (a) Adult honey bee showing the location of spiracles, the entrance to the tracheas. Tracheal mites usually live in the left-most trachea, next to the head, located in the thorax.

(b) Scanning electron microscopic view of tracheal mite. (c) Tracheal mites in the trachea.

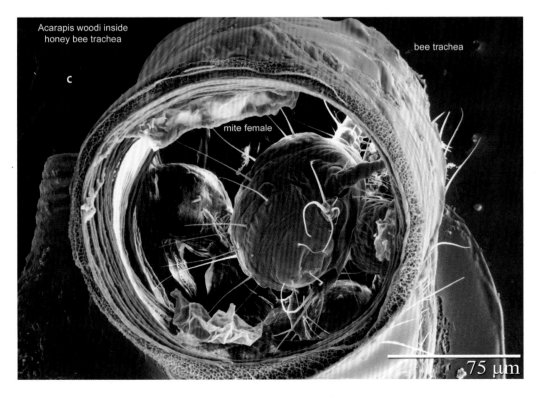

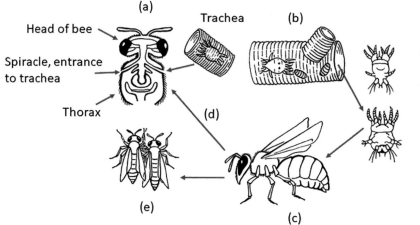

Figure 4.3: Life cycle of a tracheal mite. (a) A pregnant female mite enters the trachea via a spiracle, or entrance, on the surface of the bee. (b) Inside the trachea, the mite lays male and female eggs. Newly hatched females mate with newly hatched males. It is believed male mites can mate with sisters, since a solitary pregnant female can enter a trachea and produce female offspring that are pregnant when leaving the trachea. (c) When the pregnant female leaves the trachea, she hitches a ride on a passing bee, attaching herself to the tip of a hair of the bee. (d) When the bee that is carrying the mite passes another bee, the mite moves to the other bee and enters one of their trachea. (e) While moving around a colony, a mite attached to the tip of a bee's hair may, temporarily, move to the hair tip of another bee while waiting to infest a third bee.

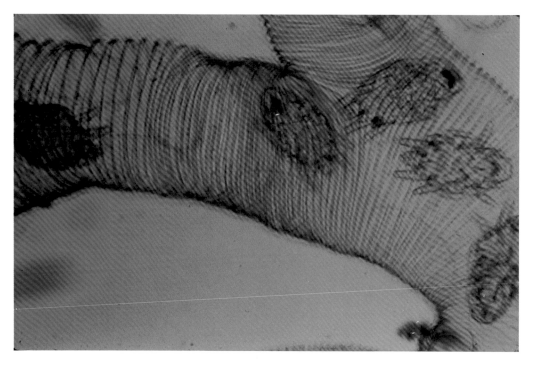

Figure 4.4: *Acarapis woodi* in a bee trachea.

Tracheae are breathing tubes that enter the abdomen and thorax of the bee at small entrances called spiracles (Figure 4.3a).

The characteristics of tracheal mite are as follows:

- They are microscopic, about 1.5 times the diameter of a human hair.
- They spend most of their life within the tracheae or breathing tubes of adult bees.
- Between three and four days after entering the host bee's trachea, each female mite lays between five and seven eggs (Figures 4.3, 4.4).
- The eggs take three to four days to hatch.
- The young adult mite leaves the host bee's trachea after five to nine days in search of a new host.

Once an adult mite has entered a bee's trachea, it will penetrate the tracheal wall with its piercing mouthparts and feed on the bee's haemolymph. The wounds eventually become surrounded by crust-like lesions (melanin patches), and the tracheae may turn nearly black. Numerous mites in varying stages of development and mite debris inside the tracheae are believed to reduce the ability of the bee to breathe, eventually killing the bees.

As part of their life cycle, mature female mites leave the bee's trachea and climb onto the tip of a body hair. When the original host bee touches a newly emerged bee, the mite will transfer to another bee where it will enter the trachea to complete its life cycle (Figure 4.3). If the mite cannot locate a new host within 24 hours of emerging, it will die.

Only female mites disperse from the host to attach to other bees, with approximately 85 per cent of the mite transfers occurring at night.

Spread

Tracheal mites are spread between colonies by bees drifting or robbing or by the swarming of infested colonies; mites are dispersed within a colony by bees brushing past one another. The susceptibility of worker honey bees to tracheal mites decreases with age, and bees over nine days old rarely become infested. Research has shown that adult female mites cannot differentiate between the ages of bees while waiting for a young host bee to pass by.

Empty equipment that has been free of live bees for one week can be reused without mite treatment.

The tracheal mite population may vary seasonally. During the period of the maximum bee population, the proportion of bees with mites is reduced.

Tracheal mite-resistant Buckfast bees

Early in the twentieth century, bee populations in the United Kingdom and Europe were decimated by the so-called Isle of Wight bee disease. This condition may have been caused by the acarine parasitic mite, as the tracheal mite is commonly called, but there is no proof.

By 1916 there were only sixteen surviving colonies in the apiary at Buckfast Abbey in Devon, United Kingdom, all of them either pure Ligurian (Italian) or of Ligurian origin, hybrids between Ligurian and the English black bee (*A. m. mellifera*). Brother Adam, who managed the apiary at Buckfast Abbey, also imported other queens, and from these, he developed what would come to be known as the Buckfast bee.

Brother Adam investigated various honey bee races and made many long journeys in Europe, Africa and the Middle East searching for pure races and attractive local stocks. Although the Buckfast bee is regarded as being of mainly Italian, or *A. mellifera ligustica*, descent, the Buckfast bee has a mixed heritage including *A. m. ligustica* (North Italian), *A. m. mellifera* (English), *A. m. mellifera* (French), *A. m. anatolica* (Turkish) and *A. m. cecropia* (Greek). The Buckfast bee of today also contains heritage from two rare and docile African stocks, *A. m. sahariensis* and *A. m. monticola*, but not the aggressive 'Africanized' *A. m. scutellata*.

To control the mating of his bees, Brother Adam moved his apiary to an isolated valley in Dartmoor, England, which became a mating station for selective breeding. Brother Adam could maintain their genetic integrity and develop desirable traits with no other bees within range.

Signs of infestation

Bees infested with tracheal mite may show one or more of several signs or even no signs at all. The mites are believed to shorten the lifespan of adult bees, affect flight activity, and may result in many bees that cannot fly, leaving them crawling in front of the hive. Studies have shown a connection between tracheal mite infestation and damage to a bee's flight muscles. The wings of infested bees are often disjointed, with the hind wing projecting 90 degrees from the axis of the body. Infested colonies may also exhibit signs of dysentery and possess an excessive inclination to leave the hive. In many cases, even severely infested colonies appear normal until their death during winter. Infested bees have also demonstrated higher bacterial levels than usual.

Treatment

The tracheal mite has become much less of a pest in the United States, but whether this is because bees have developed resistance to them or they are being controlled by *Varroa* miticides is not known.

Some British bees and some North American strains show some resistance to the mites. American bees are somewhat less resistant than British bees. Although widespread in Europe and North America, tracheal mites have not been found in Australia, New Zealand or some other parts of the world.

Acarol, Menthol, and Folbex Forte are acaricides tested in the United States, Europe and Mexico. Formic acid has been introduced under the name Apicure.

Apart from miticides, an effective method of treating the infestation is to feed colonies patties made of pollen or pollen substitute and vegetable oil. The bees brush against the patties, and their hairs get coated with vegetable oil. Once the hairs are covered with oil, it is believed the mite is either unable to hold on while waiting for a new host bee to pass, or it may impair its ability to detect young bees as hosts.

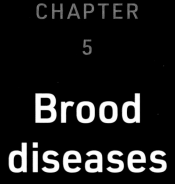

CHAPTER
5

Brood
diseases

Introduction

Most diseases of the honey bee infect brood rather than adult bees. Indeed, bees are the most susceptible to pathogens at the larval and pupal stages. Therefore, any colony inspection for signs of disease must include a rigorous examination of uncapped and capped brood cells. Although not always the case, most viral infections first display symptoms in adult bees, even if transmission occurs during their larval stage, while feeding, or from infected eggs or sperm. (Chapter 7 covers viral infections.)

As mentioned in Chapter 2, adult honey bees have several layers of defence, starting with structural barriers like a water-tight cuticle covering their outer body that blocks pathogen entry. Pathogens that are ingested will end up in the midgut, the organ responsible for digestion and absorption. Here they face another physical barrier: a chitinous substance secreted from the midgut epithelial cells that act like a sieve to block large particles such as pathogen cells from being absorbed. If pathogens breach these physical barriers, they are detected by pathogen recognition receptors which activate immune responses that may suppress or eradicate the pathogen. Young larvae, due to their still maturing midgut, leave their epithelium vulnerable to breach by bacteria and viruses. As a result, larvae are susceptible to some diseases such as American foulbrood that are fairly innocuous for adults.

Bacterial diseases of larvae

American foulbrood (AFB)

Introduction

American foulbrood, caused by the bacterium *Paenibacillus larvae*, is one of the most severe bee diseases globally. AFB can cause extensive losses in amateur and commercial apiaries. However, severe outbreaks can be minimized by good management practices, regular checks for the disease, and burning or irradiating hives when the condition is confirmed.

P. larvae spores and bacteria can remain dormant for over 50 years on equipment once used with infected colonies. *P. larvae* is contagious, although when bee larvae are more than three days old, they are less susceptible to infection.

Spread

P. larvae enters a lavae body orally and moves to the bee's gut. Once inside the gut, *P. larvae* bacteria multiply and eventually fill the space inside the gut. As bacteria fill the gut, they penetrate the gut wall and enter the larval body, infecting the haemolymph. When *P. larvae* infects the haemolymph, immune cells called macrophages engulf and destroy the AFB cells. In destroying the cells, the macrophages also die and release a poison made by the bacterium. This results in the larva dying of infection, similar to septicaemia or blood poisoning. When the larvae die, this releases possibly thousands *of P. larvae* spores, infecting many other larvae. Transmission occurs when a cell previously occupied by infected larvae is cleaned by a nurse bee, who subsequently passes bacteria to other larvae during feeding. Reuse of the contaminated cell can also spread the bacteria. Young larvae between the ages of one to four days are the most susceptible to *P. larvae*.

The control of AFB is very much in the hands of the beekeeper, as the most common method of spreading *P. larvae* is using contaminated

Figure 5.1: (a) American foulbrood. Sunken brood caps are a sign of infection. (b) The matchstick test draws out a slimy string about 3 cm (1 inch) long. If the string is shorter than this, the brood is probably infected with European foulbrood.

equipment in healthy colonies. Using infected tools and gloves, or the interchange of hive parts, or brood, within the apiary should be done with care to avoid spreading any brood disease, not just AFB. Colonies may also become infected if bees are fed honey from infected colonies, rob honey from infected colonies, or if the contaminated comb is left on the ground for other bees to feed off. Careful inspection and good hive and apiary management are the keys to preventing severe outbreaks of this disease. Studies in New Zealand show that managed colonies are more likely to be contaminated with AFB than feral colonies, highlighting that inappropriate beekeeping practices cause AFB to spread.

Signs of infection

During spring and autumn, a thorough examination of brood frames for signs of disease needs to be conducted, or at any time equipment or brood is moved between hives. It should be noted, however, that AFB can occur in hives at any time of the year. A close inspection of the frames may detect early infections, which often appear in only one or two larvae or pupae.

AFB and European foulbrood (EFB) are similar and often confused with each other, as well as with several non-disease conditions and some viral conditions. In particular, both AFB- and EFB-infected brood may have the following signs in common:

- In heavily infected colonies, the brood has a scattered, uneven pattern due to the intermingling of healthy cells with uncapped cells.
- AFB brood frames may show capped cells of dead brood with punctured and sunken capping. (Figure 5.1a)
- This peppered appearance of the brood usually allows AFB to be distinguished from EFB. With AFB, cappings are discoloured, while with EFB, cappings are not usually discoloured to any great extent.

- Both AFB and EFB can result in diseased larvae under sealed cells exhibiting a sunken dark appearance with perforated cappings.
- Brood infected with AFB generally dies after the cell has been capped. This is different from EFB, where brood generally dies before the cell is capped.
- Brood affected with AFB will change colour from a healthy pearly white to a darker brown as the disease progresses.
- Brown dead brood probed with a matchstick may show signs of a ropy consistency in infections of both AFB and EFB (Figure 5.1b). If a cell infected with AFB is tested with the matchstick test, the resulting ropiness may be 3 cm to 5 cm long (1 to 2 inches), while the ropiness test on EFB may only result in little or no stringing out of the cell contents.
- Colonies infected with AFB may have a strong, pungent smell, similar to stale beer, or have a putrid fish-like odour.
- After about a month, the brood that has died from AFB dries out to a dark scale that adheres to the cell wall. Larvae that have been killed by AFB lie on the bottom of the cell, while larvae that EFB has killed may lie in a more sideways position as if they had been wriggling out of their usual place at the time of death.
- The tongues of larvae killed by AFB are often sticking out and sometimes attached to the roof of the cell. This is not the case with larvae that EFB has killed.

The only accurate diagnosis is by laboratory analysis. Beekeepers can use a laboratory diagnostic service by submitting comb samples, honey samples, or larval smears on a glass slide. The name of laboratories in your area can be obtained by contacting your local government inspector. There are, however, easy-to-use field diagnostic kits available at a small price from beekeeping suppliers that provide reasonably accurate testing for AFB and distinguish it from EFB (Figure 5.3).

Treatment

Different countries have enacted various regulations for the control of AFB. Beekeepers need to contact a government apiary inspector, or beekeeping club, to determine which methods are recommended in their country. Treatment for colonies displaying AFB symptoms varies between countries; four of the most common methods are:

1. Kill the colony, and burn and bury the hive, or irradiate the hive with gamma radiation.
2. Treat with oxytetracycline hydrochloride (OTC) — not recommended by the authors.
3. Shaking method.
4. Administer vaccine.

Kill the colony

First, the colony of bees infected with *P. larvae* is killed with insecticide to stop infected bees from escaping or the infected comb from being robbed. Previously, the recommendation was to plug any gaps in the hive entrance, raise the lid and pour petrol inside the hive, where the bees quickly succumbed and died. This method has fallen out of favour due to the safety risks when the hive is burnt. Contaminated materials such as the hive and frames are either burnt or sterilized using gamma radiation. Both of these methods of sterilizing *P. larvae*-infected hives minimize the incidence of the disease. Fortunately, the incidence of AFB is much lower than other frequently occurring brood diseases such as European foulbrood, sacbrood or chalkbrood disease. It is usually less of a concern to beekeepers.

If an infected hive is burnt, this needs to be carried out in a pit to contain any wax and honey left over. The remaining ashes must be covered with 30 cm (1 foot) of soil to stop any residual honey from being robbed, thus spreading the infection.

Sterilizing contaminated hives using gamma radiation is also frequently used to kill bacteria and spores, although this option is only available in some areas. Before sending the infected hives for sterilization, any bees and honey need to be removed and destroyed using the same techniques as explained above. The extracted frames, boxes, hive covers, bottom boards and queen excluders are prepared for irradiation by wrapping them in a double layer of thick plastic garbage bags. After sterilization, hives can be restocked with disease-free bees.

Treat with oxytetracycline hydrochloride

The antibiotic oxytetracycline hydrochloride (OTC, trade name Terramycin) is often sold as a soluble powder and is applied in powder form to the brood area of the colony. The application of OTC in sugar syrup was practised in the past, but OTC products currently registered for control of EFB usually recommend that they are not applied as a wet treatment for bees. Wet treatment can cause residues in honey that exceed the current maximum safe residue limit. In the United States and some other countries, OTC is registered as a treatment for AFB. As noted, this is not recommended by the authors since only AFB bacteria, not spores, are killed by the antibiotic.

In countries that use OTC to manage AFB, such as the United States, beekeeping equipment may be contaminated with spores, forcing beekeepers to continuously apply this antibiotic to prevent a resurgence of the disease from spores once treatment is stopped.

Shook method

Shaking, or the shook method, is a non-antibiotic management technique in which bees are shaken from an infected frame onto new, uninfected, drawn comb. Old, infected hive material is then destroyed by burning. Results suggest that shaking bees onto frames of foundation in the spring is a feasible option for managing AFB in apiaries where antibiotic use is undesirable or prohibited. In some countries, this is not an acceptable treatment method.

Administer vaccine

The announcement of a vaccine trial for American foulbrood is an important step in the management of this and other pathogens, including viruses.

Previously, it was believed that insects, unlike vertebrates, could not acquire immunity because they lacked adaptive immunity (i.e. T and B cells) and therefore cannot produce antibodies or other specific responses that help the immune systems of invertebrates recognize and fight bacteria and viruses. Nevertheless, bees appear to have an ability for trans-generational immune priming (TGIP), whereby a queen exposed to a killed pathogen is able to produce larvae that are more resistant to this pathogen. Darlan Animal Health has developed a trial vaccine that is fed to a young queen in the bee candy that is included in every queen cage. The queen eats the vaccine and passes immunity to her offspring through the eggs in her ovaries. This is an exciting step in managing many previously untreatable bee diseases and the beekeeping industry will await results of field trials.

European foulbrood (EFB)

Introduction

European foulbrood, caused by the bacterium *Melissococcus plutonius* (syn. *M. pluton*), is a severe bee disease in some parts of the world. EFB will often cause extensive losses in both amateur and commercial apiaries. However, the beekeeper can minimize outbreaks by regular disease checks, early testing and detection, and hygienic practices to reduce the spread to other hives.

M. plutonius can remain dormant for over three years on old equipment that once housed infected colonies. EFB is highly contagious, with all stages of larval development susceptible to infection.

Spread

M. plutonius (the cause of EFB) is highly contagious, and the causes of spread are the same as for *P. larvae* (the cause of AFB). Infected colonies can survive for long periods with low spore counts without showing any severe signs. Then, sudden disease outbreaks may occur due to a build-up of bacteria numbers. Usually, these outbreaks are caused by colony stress resulting from:

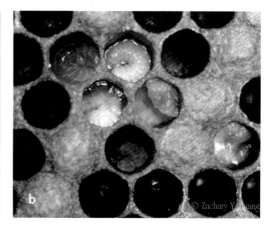
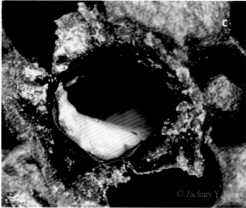

Figure 5.2: (a) Healthy larvae with no sign of disease. (b) Larvae infected with European foulbrood change colour to yellowish or brown. (c) A dead larva has dried and forms the characteristic C shape against the cell wall.

- changed seasonal conditions
- poor nutrition
- moving the bees.

Young larvae between the ages of one to four days are the most susceptible to *M. plutonius*.

M. plutonius does not attack the larva but instead enters the larva's gut and consumes food the young brood needs. The larvae are left with an insufficient amount of food on which to survive. Unlike American foulbrood, *M. plutonius* does not invade other parts of the larva's body or haemolymph but remains in the gut.

Some experienced professional beekeepers say that EFB is likely to show its presence when there is a shortage of pollen coming in. If there is an abundance of pollen and large quantities of royal jelly are being fed to the larvae, there may be sufficient food for both *M. plutonius* and the larvae so that the disease will not show itself by harming the brood.

Signs of infection

Fortunately, EFB-diseased colonies are usually easily recognized (Figure 5.2), although identification of the cause of the disease may be more difficult for the amateur due to its similarity to American foulbrood. As a rule, the following are signs of EFB:

- Brood affected with EFB may have a mottled, peppered appearance with healthy brood cells intermingled with dead or dying brood.
- Larvae are primarily affected in the unsealed, curled-up stage, although in severe cases, brood of all ages and stages of development may be affected.
- Diseased larvae collapse and become dislodged from their normal position in the cells. Their colour changes from pearly white to yellow and, finally, yellowish brown. After two to four weeks, larvae dry to form a brown scale, which can be removed from the cell.
- In some cases, capped brood is affected, and the capped brood takes on a mottled appearance with scattered sunken and perforated capping. Pupae may have a similar appearance to those affected by AFB.
- The odour of infected brood varies from odourless to sour or foul, depending on the secondary invading bacteria present.
- Outer combs of the brood nest may show signs of the disease earlier and may have a heavier infection than inner combs from the same hive.
- Dead brood probed with a matchstick usually has a watery consistency, although the sealed brown pupae may exhibit

a slight ropy consistency when a matchstick is inserted into infected larvae and then removed slowly.

- Worker bees may remove and discard diseased larvae as they die; thus, a colony may show few signs of the disease.

Diagnosis of EFB based solely on the signs described above is not always reliable. EFB can easily be confused with AFB, some viral conditions, and several non-disease conditions. The section above on AFB lists some common signs of infection for both AFB and EFB.

The only accurate diagnosis for EFB is via laboratory analysis, which beekeepers can access by submitting a comb sample, or honey sample, to a diagnostic laboratory. An EFB diagnostic kit is available similar to that used to detect AFB.

Figure 5.3: Test kits are available for both EFB and AFB. They are convenient to use in the field, even by an inexperienced beekeeper. (a) Front. (b) Back.

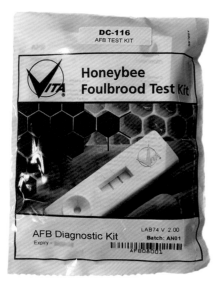
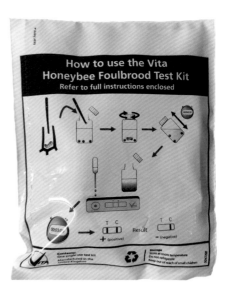

Treatment

While there may be times when antibiotic treatment would appear to be the only answer, for the hobby beekeeper, the use of antibiotics has never really been a practical solution due to the difficulty of obtaining the appropriate prescription. Their use is becoming less attractive since the honey may become contaminated, and there is the possibility that overuse will result in the development of antibiotic-resistant strains of bacteria.

The only antibiotic recommended for the treatment of EFB is oxytetracycline hydrochloride (OTC, trade name Terramycin). OTC may be available only on prescription from a veterinarian or other official, depending on the country. Some beekeepers have treated infected colonies with too much OTC, resulting in honey containing high antibiotic residues. The European Union (EU) has specified stringent limits on the amount of OTC that can be included in food-grade honey. The EU will return excessively contaminated honey to the country of origin.

Alternative methods of treatment available to the hobbyist:

- **Requeen regularly. A young, vigorous queen will always do better than an older queen, and, when buying new queens from a breeder, ensure that disease-tolerant queen breeding stock is used.**
- **Maintain hive hygiene. Regular replacement of old comb will help reduce the concentration of disease-causing organisms in the brood nest. This can be done by replacing two frames of black comb each year with two frames of white comb or foundation.**
- **Moving hives has long been recognized as being stressful to bees. Moving hives at night with an open entrance will minimize stress, as moving with a closed entrance may lead to excessive heat production and a lack of oxygen. Many apiarists report that bees are more likely to show signs of EFB soon after being moved.**

- Maintain nutrition. Nutritional problems can be divided into two categories:
  - lack of pollen
  - lack of nectar.

If ample honey is stored in the hive, a shortage of nectar should not be a problem, but good-quality pollen is another matter. Pollen is available either when the bees have stored it or when it is available from flowering plants. A good supply of pollen is essential to provide adequate amino acids to reduce any nutritional imbalance that will stress the bees. Feeding the bees either previously collected pollen or a pollen substitute will compensate for lack of pollen.

Spores and the ability of *Paenibacillus larvae* (AFB) to resist treatment

A key difference between American foulbrood and European foulbrood is the ability of *Paenibacillus larvae*, the bacterium causing AFB, to form protective spores. These spores protect the bacterium for up to 50 years, enabling it to resist treatment by antibiotics and to remain dormant in equipment ready to return to the infectious state when conditions are right. *Melissococcus plutonius*, the bacterium causing European foulbrood, is not able to form protective spores and can be treated with antibiotics such as Terramycin, which *Paenibacillus larvae* spores are susceptible to.

First, note that the correct term is endospores, not spores. Although beekeeping books regularly discuss AFB spores, a spore is a bacterium that can divide and reproduce. A spore is often unable to protect bacteria the way an endospore can. In this book, we will follow tradition and use the term spore to mean the protective form of the bacterium causing AFB.

Cell duplication (meiosis)

Bacteria such as *Paenibacillus larvae* and *Melissococcus plutonius* reproduce by the process of cell duplication called meiosis. With meiosis, a single cell

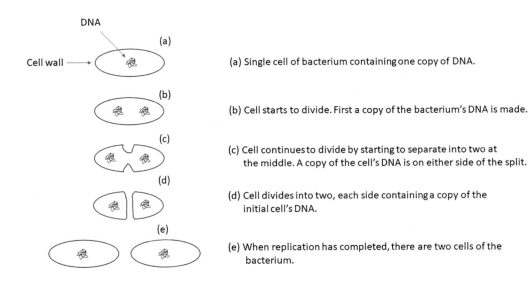

Figure 5.4: **Division of a cell by meiosis.**

divides in two (Figure 5.4). Before dividing, the cell makes a copy of its DNA (b); the single cell divides to produce two cells with identical DNA (c) and (d); the result is two cells that contain the same DNA and which are able to replicate again to form four cells.

Formation of endospores by *Paenibacillus larvae*

As mentioned, meiosis is different from the formation of endospores. Meiosis involves increasing the number of viable bacterial cells. The formation of the endospore involves replacing the weaker cell wall enclosing the DNA with a tougher protective shield during times of stress, such as a lack of nutrients. *Paenibacillus larvae*, the causative agent of AFB, forms endospores, while *Melissococcus plutonius*, the causative agent of EFB, does not.

The *Paenibacillus larvae* bacterial cell, sensing that the environment is hostile because, say, there is a lack of food, starts the process to form an endospore, which is a form of hibernation (Figure 5.5). First, the cell produces a copy of its DNA and forms a wall that divides the cell into two (b). Next, a protective membrane starts to form around the DNA on the right,

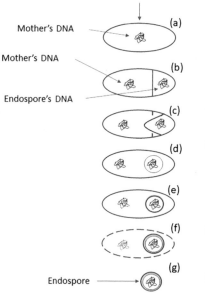

Mother vegetative cell

Mother's DNA (a)

(a) Single cell of *Paenibacillus larvae* starved of nutrients

Mother's DNA (b)

Endospore's DNA

(b) DNA is duplicated,
Cell divides in two within itself,
Wall forms between two sets of DNA,
Smaller compartment on right.

(c)

(c) Wall breaks down,
Membrane starts to form around endospore's DNA on right.

(d)

(d) Membrane forms around endospore's DNA,
Endospore DNA enclosed by simple membrane,
Endospore DNA floats in mother vegetative cell.

(e)

(e) Wall of endospore membrane thickens to provide further protection,
Mother's DNA breaks down.

(f)

(f) Mother cell wall breaks down,
Endospore breaks free from mother cell,
Endospore forms three protective layers in shell.

(g)

Endospore

(g) *Paenibacillus larvae* endospore able to survive in a hostile environment

Figure 5.5: **Formation of endospores by *Paenibacillus larvae*.**

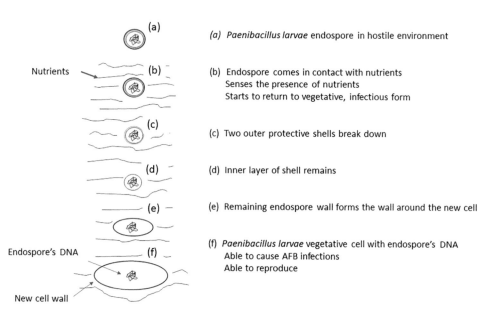

(a)

(a) *Paenibacillus larvae* endospore in hostile environment

Nutrients (b)

(b) Endospore comes in contact with nutrients
Senses the presence of nutrients
Starts to return to vegetative, infectious form

(c)

(c) Two outer protective shells break down

(d)

(d) Inner layer of shell remains

(e)

(e) Remaining endospore wall forms the wall around the new cell

Endospore's DNA (f)

New cell wall

(f) *Paenibacillus larvae* vegetative cell with endospore's DNA
Able to cause AFB infections
Able to reproduce

Figure 5.6: **Return of *P. larvae* endospore to the infectious state.**

and the internal wall dividing the cell breaks down (c). The membrane protecting the new DNA thickens, providing additional protection, while the original DNA breaks down (e). The original cell wall breaks down, and a third protective layer is added to the endospore (f) and (g). The process of endospore formation is now complete, and the three layers protect the DNA from antibiotics, other harmful chemicals or dehydration. The newly formed *Paenibacillus larvae* endospore can remain protected, and dormant, for 50 or more years before returning to an active state.

Breakdown of *Paenibacillus larvae* endospores

Figure 5.6 shows the steps taken by a *P. larvae* endospore to return to its infectious state when it senses that it is no longer in a hostile environment. The endospore detects, say, the presence of nutrients and starts the process to return to the vegetative state (b). First, the two outer layers of its protective shell break down, leaving only the inner protective layer (c) and (d). The remaining inner protective layer, or shell, forms the cell wall of the vegetative cell (e). The endospore completes its transformation into an active, infectious, vegetative cell, able to cause AFB, and capable of reproducing by meiosis.

Fungal diseases of larvae

Chalkbrood

Introduction

Chalkbrood is not usually a severe disease of honey bees and rarely kills an entire colony. If the bees are susceptible to the disease, it can cause a gradual deterioration of the colony by killing brood, affecting adult bee numbers and resulting in a loss of honey production. Chalkbrood is caused by the fungus *Ascosphaera apis*, and there are signs that its incidence is increasing worldwide, probably due in part to migratory beekeeping practices and

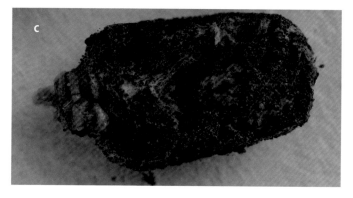

Figure 5.7: (a) Mummies infected with chalkbrood at entrance to hive. (b) Brood killed by chalkbrood inside cells. (c) Mummy from front of hive.

poor hygiene by some beekeepers. The fungus can remain active on hives and other surfaces for up to fifteen years. There has also been the suggestion that importing infected pollen and foundation may be an additional cause of the increased disease incidence.

Both *Ascosphaera apis* and mushrooms are types of fungi. The white and black *A. apis*-infested bee larvae cadavers can be compared with parts of a common mushroom (Figure 5.8b). The body of a mushroom consists of filaments called hyphae bundled together to form the solid structure of the mushroom, the stalk, the cap, as well as underground hyphae, similar to roots. The filaments that constitute a white *A. apis*-infected larva are also hyphae. Later in the development of a chalkbrood infection, the ends of the filaments turn grey or black. The grey-black form of a chalkbrood infection is the development of vegetative spores, used by *A. apis* for reproduction

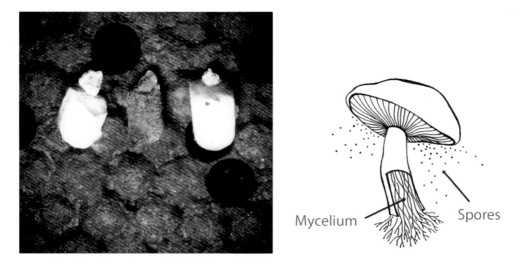

Figure 5.8: **Comparison of a mushroom fungus with white and grey/black mummies infected by the chalkbrood fungus *Ascosphaera apis*. (a) White and grey/black mummies displaying chalkbrood. (b) A typical mushroom. The white chalkbrood is the mycelium, which provides the roots and body of the fungus. The black chalkbrood are the spores, which are the reproductive, or vegetative, spores of chalkbrood. The white mycelium grows first, followed later by black reproductive spores.**

(Figure 5.8a). These vegetative spores are similar to the spores released by mushrooms to reproduce.

Spread

Chalkbrood fungus (*Ascosphaera apis*) can be passed to an uninfected hive in several ways: by forager bees returning with infected pollen or nectar, robber bees stealing honey, or by the drift of bees. The fungus is passed orally to uninfected bees through contaminated food or worker bees cleaning the inside of an infected colony, causing the fungus to be consumed. While adult bees are not susceptible to chalkbrood disease, they are the carriers and pass the fungus to larvae that consume it, infecting the larvae, which may eventually die.

Young larvae are the most susceptible to the disease between the ages of one to four days. When chalkbrood fungus has been consumed by the larvae and is inside its gut, the fungus penetrates the gut wall and grows inside the body cavity of the larvae. As the infection develops the fungus eventually breaks out of the body cavity, penetrates the outer surface or skin, initially at the anal end but eventually breaking out over the body, covering the larvae with a thick white layer of mycelium. Mycelium is similar to household mould on fruit and bread. At the end of each string of mould on the outside of a black cadaver there will be a new infectious spore, with as many as one hundred million to one billion active spores on the outside of each black cadaver ready to be ingested by unsuspecting adult bees.

Young infected larvae do not usually show signs of disease but will die at the late larval, or pre-pupal, stage, after being sealed in their cells. Worker bees uncap the cells of dead larvae, making mummies visible, before sometimes removing the mummified pupae and depositing them on the hive floor or at the entrance to the hive.

Honey bee larvae that have died from chalkbrood infection are initially swollen in size, taking up the entire volume in the cell. Eventually, the carcass dries out, leaving the characteristic chalk-like pellet associated with the disease.

Signs of infection

Chalkbrood can infect both unsealed larvae and sealed pupae; in this respect, it is similar to sacbrood. The main signs of the disease:

- Infected brood are called mummies (Figure 5.7b). When mummies are removed from their cell they appear to be solid, similar to lumps of chalk.
- Partially chewed-down mummies are often seen in open cells.
- Mummies vary in colour from white to dark grey or black.
- Many of the sealed cell caps may have a small pinprick-sized hole.
- Dead larvae will be dropped onto the hive floor by nursery bees and later moved outside the hive entrance by worker bees, leaving a litter of chalk-like mummies on the ground (Figures 5.7a and c).
- A hive heavily infected with chalkbrood will show a patchy, scattered brood pattern.
- Chalkbrood is more prevalent during the spring since fungal growth increases during cool, damp conditions, particularly in poorly ventilated hives.
- Chalkbrood can infect the larvae of workers, drones or queens.

Treatment

There are no chemical treatments for chalkbrood, so, like many other honey bee diseases, good management practices by the beekeeper lie at the front line of control and cure. These include:

- Requeening frequently with queens bred by a reputable queen rearer who monitors colonies closely and selectively rejects queens from colonies having chalkbrood during the season.

- Use of bees that have originated from hygienic stock that are resistant to the disease appears to be the optimum method of clearing up signs of the disease.
- Hives should be well ventilated and free from damp.
- Beekeepers need to maintain a high level of hygiene by cleaning used equipment, mainly the hive tool and their gloves, before moving on to inspect other hives.
- Studies have shown that using old brood comb harbouring the disease can cause further disease outbreaks. The beekeeper needs to implement a rotation system, replacing two to three old brood combs yearly with new frames and foundation.

Minimize colony stress as the disease occurs most frequently in colonies that are expanding during the late spring and summer:

- Feed syrup and pollen or pollen substitute during periods when there are few flowers available for food.
- Keep movement of the hive to a minimum.
- Avoid the transfer of frames or other hive parts between colonies.

Some beekeepers shake brood frames; if chalkbrood is present in large numbers, the rattling of dead larvae may be heard.

Stonebrood

Introduction

Stonebrood, caused by either of the two fungi *Aspergillus flavus* or *Aspergillus fumigatus*, is rarely detected in colonies, so its impact on bee health is poorly understood. The fungi are commonly found in the soil and are harmful to other insects, birds and mammals (including humans), as well as some plants. Stonebrood attacks the brood and transforms the larvae into a hard, stone-like, coloured (green, yellow, grey or black) object found lying in open cells. Adult bees may also become infected and killed by the fungi.

Stonebrood is often discussed with chalkbrood, although they are not the same. A bee infected with stonebrood fungus may die within as little as two to four days after infection. *Aspergillus* fungi are pathogens that require an immunocompromised host to establish infection. Susceptibility to stonebrood is believed to be increased by dampness; chilling of the brood; weakening of the colony by other diseases such as viruses, *Varroa*, or foulbrood; manipulations that decrease the brood to adult bee ratio; and poor hygienic behaviour of nurse bees.

Stonebrood may be identified by inspection of the dead larvae. Larvae killed by stonebrood are hard, hence the name, and are greenish or yellowish. The dead larvae's colour may reflect the infection's cause since *Aspergillus flavus* is yellow-green while *A. fumigatus* is grey-green. If an adult bee is affected by either of the fungi, soon after death, the abdomen of the infected bee may become hard, which may be the only evidence of the distinctive symptom of the disease. The fungus can also grow on the surface of a bee's body, causing damage from the outside.

Treatment

Due to its infrequency, stonebrood is not considered a serious disease of the honey bee and is best controlled using the same management practices as for sacbrood (see Chapter 7: Viruses).

Adult diseases

This chapter discusses *Nosema apis* and *Nosema ceranae*, as well as bacterial diseases caused by *Spiroplasma*. Viral infections of adult bees are covered in Chapter 7.

x

Nosemosis

Introduction

The most widespread pathogens related to the global decline in honey bee health, after *Varroa*, are *Nosema apis* and *Nosema ceranae*. These small parasites live in the digestive tract of honey bees. Infection by either *N. apis* or *N. ceranae* is called nosemosis. Historically, European honey bees were only infected with *N. apis*, called Type A (Apis) nosemosis. In 1996, a second species, *N. ceranae*, was reported from Asia to have jumped from the Asian honey bee, *Apis cerana*, to the European honey bee, *A. mellifera*, called Type C (Ceranae) nosemosis. Initially restricted to Asia, *N. ceranae* quickly spread globally and is now regarded as the dominant species of *Nosema*.

Both *Nosema* species are microsporidians, fungus-like pathogens that live and reproduce in the bee's gut. Microsporidians have a more basic genetic structure than most pathogens and cannot manufacture the chemical adenosine triphosphate (ATP), which is used by all animal cells to provide energy for cell function. As a result, *Nosema* needs to use ATP manufactured by cells in the bee's gut lining to supply it with energy, reducing the lifespan of individual bees, sometimes leading to colony collapse.

Spread

Bees become infected with both species of *Nosema* through the mouth. Infection can occur when the adult bee is feeding on infected honey, or when adult bees clean previously contaminated cells. Larvae can become infected from previously infected cells, or when they are fed infected royal jelly or bee bread.

When spores of *Nosema* enter the gut, they germinate and penetrate the lining or wall of the midgut, the epithelial layer, where they multiply rapidly

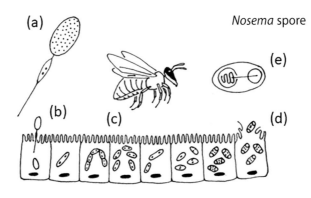

(a)

Nosema spore

(e)

(b)

(c)

(d)

Epithelium, or gut wall, of bee

Figure 6.1: **Life cycle of *Nosema* in the gut. (a) Within 30 minutes of entering the gut, the *Nosema* spore germinates and everts its polar tubule, a hollow needle-like structure. (b) The tubule comes into contact with the epithelium, or gut wall, penetrates the cell wall, and injects infective plasma into the cell. (c) Inside the cell, the plasma rapidly multiplies, forming *Nosema* spores. This phase lasts six to ten days and uses the cell contents as its food supply. (d) When the cell is full of mature spores, the cell wall ruptures, releasing spores into the gut. (e) Newly released spores either enter new cells on the gut wall or are excreted through faeces, to infect other bees in the colony. The tubule is shown retracted in (e).**

(Figure 6.1). New spores are formed in the epithelial cells. The cells rupture and shed spores into the midgut cavity where they either go on to infect other cells in the gut wall or are passed into the faeces of the bee. The cycle begins again when other bees take in spores, such as when eating contaminated food or cleaning the colony. Spores can remain viable for many months in dried faeces on brood comb. Towards the end of winter, if adult bees cannot leave the colony, combs are often soiled with faeces from infected workers. Other bees become infected when they pick up the spores in the faeces as they clean the soiled combs during the spring expansion of the brood nest.

Unlike other animals, epithelial cells of the honey bee do not secrete digestive fluids into the gut. Instead, the bee's epithelial cells break free

Figure 6.2: *Nosema apis* and *N. ceranae* spores observed under a scanning electron microscope. (a) The lining of the gut is covered with *Nosema apis* spores. (b) *Nosema apis* spore. (c) Arrows indicate the start of the extrusion of polar tubules. Tubules are used by the spores to penetrate the gut cell lining and are used to inject *Nosema* DNA into the host cell. There is little difference in the shape of the two spores.

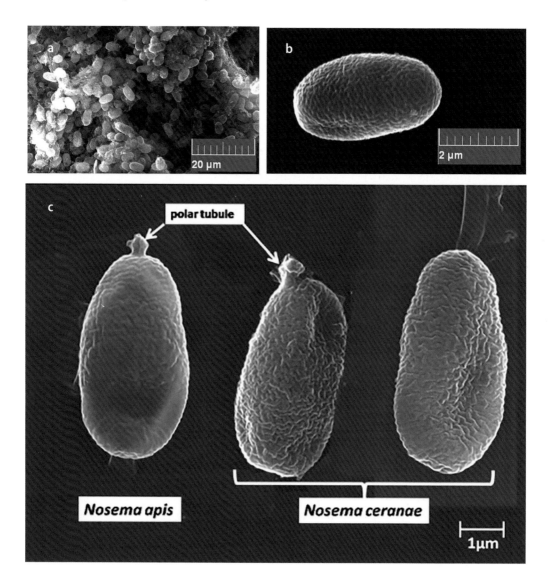

from the gut wall and burst, releasing enzymes into the gut to be used by the adult bee to digest food. If the cells lining the gut wall, or epithelium, contain N. apis spores, the burst cell will release spores into the gut, where they either repeat the cycle or are excreted to infect other nurse bees that are cleaning faeces from the hive.

Figure 6.2 shows N. apis and N. ceranae spores. The main difference is that N. ceranae spores have a much rougher surface than N. apis (the rougher surface is clearer in Figure 6.2b). In one study, where the surface of the gut or intestines was studied with a scanning electron microscope (SEM), spores were present in such a large number that they formed a uniform layer on the surface of the midgut (Figure 6.2a), which could be the cause of the decline or loss of a honey bee's midgut function. The spore layer covering the midgut's lining (the epithelium), even during medium infestations, can lead to bee malnutrition and may be the cause of higher mortality among bees observed after infection with Nosema. If the lining of the midgut is covered with spores and the bee is unable to absorb sufficient nutrients and energy from digested food, activities that are heavily dependent on a bee's ability to store and expend energy, such as foraging, will be impaired.

Signs of infestation

There are no symptoms of the disease that are unique to nosemosis. The inability of bees to fly, faeces on combs or lighting boards, and dead or dying bees on the ground in front of the hive may indicate infection by Nosema, but they may also be caused by other pathogens or abnormal conditions. N. apis may cause the digestive cavity part of the midgut of heavily infected bees to become white, soft and swollen, while N. ceranae infections do not. Wings spread in the shape of a K may signify Nosema infection (Figure 6.3).

N. apis and N. ceranae possess a similar internal structure but differ in the size of the spores, with the spores of N. apis being larger than those of N. ceranae (6×3 μm versus 4×2.2 μm). Nosemosis is challenging to

Figure 6.3: **Wings spread in the shape of a K may signify *Nosema* infection. Other pathogens can also cause K-wings, including some viruses. If you see many bees in a colony with K-wings, they may be infected, but this symptom alone is not enough to identify the disease.**

diagnose; a microscopic examination of the crushed abdomens of about 60 bees, mixed with a small amount of distilled water, is the only reliable test for the presence of this disease. Both species of *Nosema* are visible under the microscope, although differentiating between the two species is only possible using genetic PCR testing.

Infection with *N. apis*, Type A nosemosis, usually occurs with colonies that are already weakened during extended periods of poor weather or when the colony is confined within the hive. Some commercial beekeeping practices also favour the development of nosemosis, and, in the United

States, the pathogen is more prevalent in commercial colonies than in colonies kept by hobbyists. On the other hand, in the United States *Varroa* is more commonplace than *Nosema* in colonies held by hobbyists.

Infection with *N. ceranae*, Type C nosemosis, may result in no signs of infection or significant colony health problems and a resulting economic loss to the beekeeper. Some studies have found a relationship between infection with this pathogen and the collapse of a colony. *N. ceranae* shows a seasonal pattern of infection directly related to increased temperature. When more than 80 per cent of the bees in a colony are infected with more than 10 million spores each, colony collapse occurs.

Mixed infections of both pathogens occur, although *N. apis* has been replaced by *N. ceranae* in tropical areas. The coexistence of both species in colonies has also been reported in cold areas, even though the viability of *N. ceranae* spores decreases by around 30 per cent at low temperatures.

Both species of *Nosema* are mainly transmitted by the oral-faecal route, where an adult bee becomes infected by cleaning infected cells. Infection through trophallaxis, when an adult bee feeds another bee, is also possible. Feeding larvae contaminated with royal jelly, pollen or bee bread causes infection in the brood. Although much less critical to disease spread, vertical transmission from the queen to her offspring is also possible since *N. ceranae* spores have been found in the ovaries of queens. Beekeepers may also aid the spread between colonies since the pathogen has been detected in wax, honey and pollen.

Infection with *N. ceranae* reduces the ability of bees to fly and causes disorientation during flight, possibly due to the removal of ATP by the pathogen. *N. ceranae* can also affect colony behaviour with lower honey and offspring production and lower numbers of worker bees in infected colonies. The use of insecticides is a significant factor causing synergies with *N. ceranae* infection.

As discussed in Chapter 2: How Pathogens Infect Bees, besides individual immunity, at the colony level, bees use a defence mechanism called social immunity, which includes propolis production. Social immunity also includes corpse removal, self-removal of sick individuals, grooming and

Figure 6.4: **Diarrhoea on the front of a hive is not a reliable indicator of** *Nosema.*

fever — increasing the body temperature around a pathogen — and secretion of antimicrobial molecules in food by nurses for young bees. Since grooming involves licking and chewing, this may favour the spread of *Nosema*. Also, bees infected with *Nosema* are known to be weaker at flying, have impaired navigation, and often die before returning to the hive; these may be adaptations to lower the infection rate within a colony.

The ability to resist infection also depends on external factors, such as the food resources available, which are essential to compensate for the removal of ATP and to boost the bee's immune system.

Diarrhoea is not a reliable indicator of nosemosis (Figure 6.4). Diarrhoea may be caused by *Nosema*, it may be absent in an infected colony or may be the result of some other cause such as diet.

Treatment

Fumagillin, a mycotoxin, is one of the most common treatments, which can be administered either as a prophylactic or control treatment. Both *N. apis* and *N. ceranae* are sensitive to fumagillin, which temporarily reduces nosemosis and the risk of collapse. However, studies show that the size of treated and non-treated colonies was similar two months after treatment, so the probability of surviving the winter, with or without treatment, does not differ between them.

For treatment, fumagillin is dissolved in sugar syrup and fed to the colony, usually in the early spring, before *Nosema* has increased sufficiently to harm the colony. Treatment should end four to six weeks before honey supers are added to minimize the risk of contamination. In the United States, the recommended dose of fumagillin is greater for *N. ceranae* than for *N. apis*. When deciding how much to administer, beekeepers should obtain advice from a local government apiarist, beekeeping supplier or bee club.

On the negative side, fumagillin seems to impair normal cell function in the bee. There are also signs that *Nosema* is becoming resistant to fumagillin, a severe problem with all insecticide and antibiotic treatments. In the United States, fumagillin is the only antibiotic approved for the control of nosemosis; its use, however, is banned in Europe and Australia due to the possible presence of residues in honey. Research on alternative treatments has become urgent. Combining different treatments could be an excellent way to diminish bee losses due to *N. ceranae*. However, other actions, such as improving the management of colonies and reducing insecticides, are necessary to increase bee populations worldwide. The role of integrated pest management, as with other bee pathogens, is critical in this regard.

Spiroplasma

Introduction

Two species of the bacterium *Spiroplasma*, *S. apis* and *S. mellifera*, have been found to infect honey bees, and their presence in a colony is associated with health issues, including the death of individual bees, but not colony death. *Spiroplasma* is a class of bacteria whose significance as pathogens of the honey bee has only recently been established. Members of the species were first identified as infecting plants, but it is with arthropods and insects, including bees, that infection has the greatest health significance.

The way *S. apis* and *S. mellifera* infect bees has not been elucidated. However, its harmful effect is believed to depend on its ability to move out of the midgut, through the epithelium (the lining of the gut) and into the haemolymph circulating in the bee's body. *Spiroplasma* may also infect other host tissue.

S. apis was first identified in France, and found to be associated with 'May disease', a malady usually experienced in May or June in European countries. Other studies, however, have not shown an infection peak in May/June, and the bacterium is found evenly throughout the year.

Signs of infestation

Bees infected with *S. apis* or *S. mellifera* are unable to fly, can only crawl on the ground, and their intestine is filled with undigested pollen. The main visible effect of the pathogen on individual bees is neurological disorders, such as quivering, creeping along the ground, and moving in small groups away from the colony where they die. Colonies infected with the bacterium can show a greater than 25 per cent reduction in honey production. Like most other infections of adult bees, the pathogen enters the bee orally, resulting in reduced bee lifespan.

Although *Spiroplasma* in honey bees was reported in 1976, scientists have only recently investigated the role the bacterium plays in bee and colony health. Most previous research involved determining the geographic spread of the pathogen; for instance, it is known to infect Africanized bees in Brazil.

Treatment

Unlike the two other bacterial pathogens in honey bees, *Paenibacillus larvae* (AFB) and *Melissococcus plutonius* (EFB), *Spiroplasma*-infected bees are impossible to identify by outward pathology. Colonies usually recover spontaneously from the disease without treatment.

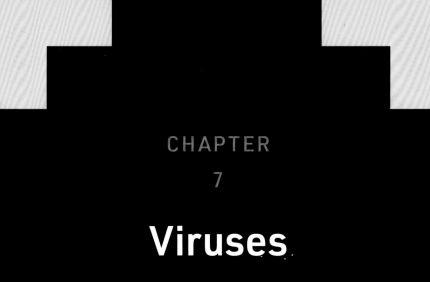

CHAPTER
7

Viruses

Common viral diseases

Viruses can infect all life forms; some are benign and cause no known harm to bees or other animals, while others can be lethal. Although some animal species may have common viruses, some viruses can survive only in a single host, such as a honey bee. Viruses cannot live outside of another organism's cell and can only breed and multiply in its host cells.

Bees have been infected with viruses for possibly thousands of years. During this time, they have developed a range of effective defences against them. *Varroa*, however, changed all this by transmitting lethal viruses directly into haemolymph, bypassing the effective immune response that had protected bees for many years. The following is a summary of the more common viruses found to infect honey bees. There are many more, although the majority constitute little threat to bees.

Viral diseases of the larvae

Sacbrood virus (SBV)

Sacbrood virus, a rapidly breeding Iflavirus, infects larvae, which then typically die after capping but before the change to the pupal stage of their development in the capped cell. Infected larvae form a plastic-like outer layer or skin containing a watery fluid, giving the larvae its distinctive bloated look. Infections are most common during the first half of the brood-rearing season when there is a relatively rapid turnover of the hive population from the long-lived 'winter bees' to a new generation of short-lived 'summer bees'. The colony at this time is very susceptible to both chill and nutritional stress, which can lead to further progression of the sacbrood disease or the development of other viral infections.

Even though hygienic adult bees can detect and remove dead larvae quickly, there is evidence that some non-hygienic bees do not remove infected larvae. This causes the queen to miss laying a new egg in the old cells infected by dead larvae, resulting in a patchy distribution of new brood.

In most apiaries, sacbrood is limited to only one or two hives. If, however, the infection is sufficiently developed in a hive to alert the beekeeper to the signs, the disease may be severe enough to cause the numbers of the adult worker population to be affected.

Signs of infection

Sacbrood mainly affects female worker larvae, although it may sometimes occur in male drones. The disease affects uncapped larvae and sealed brood that are seven to ten days old. Brood comb contaminated with sacbrood show some degree of irregularity of brood pattern similar to EFB or AFB. Dead larvae change colour from a white to a yellow and then brown (Figure 7.1). Dead brood may be scattered among healthy brood in the comb, and their capping may be discoloured, sunken, perforated, or removed by the bees.

Dried larvae that have died of sacbrood are located on the bottom of

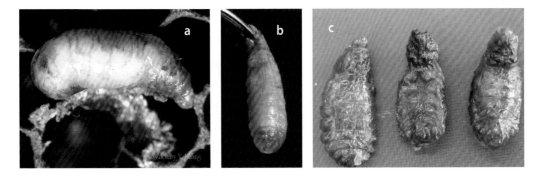

Figure 7.1: (a) Larva infected with sacbrood virus. (b) SBV-infected larva showing a watery sac at the bottom. (c) Larvae killed by SBV with darkened heads.

the cell and retain the upturned 'gondola' shape of the body. Over time, the dead larva dries out, becomes brittle and turns brown to black. The dried scale can easily be removed without damaging the cell wall. Infected brood is typically odourless, although, in advanced stages, an occasional putrid smell may be detected.

Typical signs of this disease:

- Dark, upturned heads of the dead larvae are the most visible sign in open cells.
- Most larvae killed by SBV have dark, reddish-brown head areas with less dark, straw-coloured thoracic and abdominal areas. AFB-infected larvae have a dark coffee colour.
- Darkening begins at the head and then spreads to the rest of the body.
- The skin of a dead larva changes into a tough, plastic-like sac. Between the skin and the larval body, a greyish granular fluid accumulates.
- Capped cells containing infected larvae often have a pin-prick-sized hole in the capping.
- Body segmentation is maintained in the dead larvae.
- Unlike EFB or AFB, no ropiness is associated with the matchstick test of the moist dead larva. The rope test used in the field determines whether the larva has died of sacbrood, EFB or AFB.
- EFB may be mistaken for sacbrood since the signs of both infections are similar (Figure 7.2).

Treatment

The incidence of sacbrood in most colonies is low because adult bees usually detect and quickly remove infected larvae. Good management practices by the beekeeper that can help alleviate sacbrood include:

- Requeening with a young, vigorous queen.

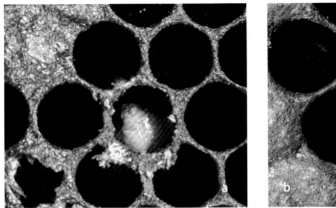

Figure 7.2: **Comparison of sacbrood virus and European foulbrood symptoms. SBV and EFB can look similar and are difficult to tell apart. (a) Larva infected with SBV. (b) Larva infected with EFB.**

- Strengthening any weakened hive with a disease-free nucleus colony (nuc) of young bees.
- Moving bees to better conditions with a good honey flow.
- Feeding bees a sugar solution and provide good-quality pollen or pollen substitute if natural supplies are low.
- Packing the bees down into fewer boxes.
- Routinely replacing old brood combs every year.

Spread

- Nursery bees probably become infected when cleaning out the cells of infected pupae and then passing the virus on to non-infected larvae where the virus breeds rapidly, leading to death shortly after capping.
- In adults, sacbrood accumulates in the hypopharyngeal glands of nurse bees without appearing to cause any signs of the disease.
- Since the hypopharyngeal gland is used in the production of royal jelly and brood food, it is believed this is a path for larvae to become infected.
- The virus may remain active for up to four weeks in larval remains or in honey or pollen.

Black queen cell virus (BQCV)

Black queen cell virus, a common infection of adult bees, is believed to be the most significant cause of queen pupae death in Australia and North America. The term black queen cell virus comes from a symptom of the disease where the infected queen larvae and pupae turn black due to the formation of melanin, which is then absorbed by the wax walls of the cell and become visible on the outside of the cell (Figure 7.3). In Europe, however, where about 50 per cent of larva are infected with BQCV, larvae do not turn black but remain a creamy colour. The walls of the cell similarly do not turn black.

Infection of adult bees with BQCV and nosemosis are linked, and it is *Nosema* spores within the gut that enable BQCV to enter the adult bee's body initially. Larvae become infected when nurse bees with BQCV feed the larvae.

Signs of infection

The characteristic signs of BQCV are that dead queen larvae in open cells

Figure 7.3: (left) Queen cells coloured black by BQCV-infected larvae.
Figure 7.4: (right) Larva infected with black queen cell virus. (a) Dead larva coloured yellow. (b) Dead larva with darkened head.

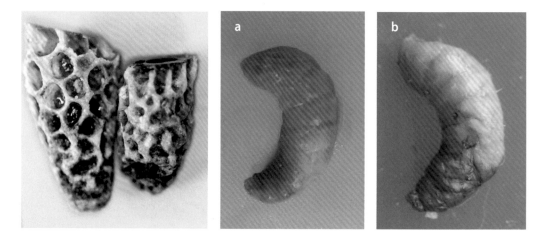

or dead pupae in sealed cells have turned yellowish and then their colour has progressed to dark brown and black (Figure 7.4). Although the virus is called black queen cell virus, it can also infect worker and drone pupae, without showing any visible signs, or the affected pupae may look like it has been killed by sacbrood virus.

Treatment

There is no treatment for BQCV. The following integrated pest management techniques may slow the spread of the virus: sterilization of grafting tools (in ethanol or by flame); control of *Varroa* and *Nosema*; and well-fed breeder, starters, cell builders, and mother colonies when rearing queens.

Some beekeepers report that antimicrobials like fumagillin or Terramycin can clear up BQCV symptoms. This has not been proven, although it may be possible if it disrupts interaction with *Nosema*.

Viral diseases of the adult bee

The list below details the six most crucial adult honey bee viruses:

1. Deformed wing virus (DWV)
2. Israeli acute paralysis virus (IAPV)
3. Acute bee paralysis virus (ABPV)
4. Kashmir bee virus (KBV)
5. Chronic bee paralysis virus (CBPV)
6. Lake Sinai viruses (LSV).

Deformed wing virus (DWV)

DWV, of which there are two major strains, is one of a few bee viruses easily recognizable due to its well-defined signs in some infected bees. Typical symptoms of DWV infection include:

- stubby, useless wings (Figure 7.5)
- shortened, rounded abdomens
- discolouring of the adult bee's body
- paralysis.

In addition to the above more apparent signs, DWV-infected bees have impaired learning capabilities, although this deficiency is not evident outside of laboratory conditions. DWV is one of the most prevalent infections in honey bees and is known to be transferred by *Varroa destructor*. Studies of *V. destructor* have shown that up to 100 per cent of the mites are carrying the virus.

- DWV tends to remain at low levels in healthy colonies and may exist as a continuing low-level infection with no visible signs.
- If the colony becomes stressed, the viral load increases and, as a result, adults emerge from the cell with deformities proportionate to the scale of their viral load.
- If this loss is excessive and can no longer be compensated for by the emergence of new healthy bees, the colony rapidly dwindles and dies.

Studies have shown that both DWV and acute bee paralysis virus generally follow the rise and fall of *Varroa* infestation in the hive and peak in late summer and early autumn.

Kakugo virus is closely related to DWV but is only found in guard bees where it increases their level of aggression.

Israeli acute paralysis virus (IAPV)

Acute bee paralysis virus Kashmir bee virus, and Israeli acute paralysis virus are part of a complex of closely related viruses from the *Dicistroviridae* family. This close relationship probably indicates a common and recent

Figure 7.5: **Worker showing signs of deformed wing virus.**

ancestry for these viruses. *Dicistroviridae* is a family of viruses that infect many species of invertebrates, including aphids, leafhoppers, flies, bees, ants and silkworms.

Israeli acute paralysis virus was named after its first discovery in Israel in 2004. Some researchers believe the *Varroa* mite transmits IAPV, although this method of transmission is in dispute. The time of peak infection with IAPV is unknown.

Infected honey bees may have signs including shivering wings, darkened and hairless thorax and abdomen, and progressive paralysis followed by death. These signs are similar to those of both acute bee paralysis virus and Kashmir bee virus, for which it may be mistaken. IAPV affects all castes of the honey bee during all stages of their life: queens, workers and drones during their larval, pupal and adult phases can all be infected by the disease.

Acute bee paralysis virus (ABPV)

Since its first identification in the European honey bee in the early 1960s, acute bee paralysis virus has been reported in honey bees globally. The spread of ABPV in colonies is believed to occur through the salivary gland secretion of infected adult bees when those secretions, such as royal jelly, are fed to young larvae. ABPV accumulates in adult bees' hypopharyngeal glands, which produce royal jelly to feed the developing brood. If the brood is infected with large numbers of viral particles, the larvae will die before they are sealed in brood cells. If the larvae eat less lethal amounts of viral particles, they will likely survive and emerge without any apparent signs of infection.

Infection with ABPV tracks *Varroa* infestation levels and peaks generally in late summer and autumn. Similar to both Kashmir bee virus and Israeli acute paralysis virus, the virus can induce early trembling signs and rapid mortality, sometimes leading to the death of the brood. Although the symptoms of ABPV are similar to those of chronic bee paralysis virus, the viruses are different, with ABPV being the more lethal of the two.

Kashmir bee virus (KBV)

First discovered in the Asian bee, *A. cerana*, in 1977, Kashmir bee virus was later identified in the European honey bee. The virus attacks all stages of the bee life cycle and may be present in brood and adult bees without the bee showing any apparent signs. Among all of the viruses infecting honey bees, KBV is considered to be highly infectious and multiplies quickly once a few viral particles are introduced into a bee's haemolymph. KBV can cause bee mortality within three days of infection. *Varroa* mites can efficiently transmit the virus, but the parasitic mites may also activate a latent infection already within the bee.

Although KBV was first found in the Asian honey bee *A. cerana*, KBV was subsequently found in Australia in the 1970s and elsewhere without *A. cerana*. There is no good evidence that KBV spread from *A. cerana* to *A. mellifera*.

Chronic paralysis virus (CBPV)

Bees, globally, are infected with chronic bee paralysis virus (Figures 7.6 and 7.7). There are two methods of infection that have been proposed:

1. Larvae may become infected via salivary gland secretions of infected adult bees when these secretions are fed mixed with pollen as bee bread. Larvae infected with large numbers of CBPV particles die before they are sealed in brood cells. If, however, the larvae have lower infection levels, the adult may emerge, as in the case of ABPV, with no apparent signs of the disease.

Figure 7.6: Adult with (form 1) CBPV, showing the alimentary tract and thorax. The arrow shows an expanded crop due to excess food. Spread wings are also depicted.

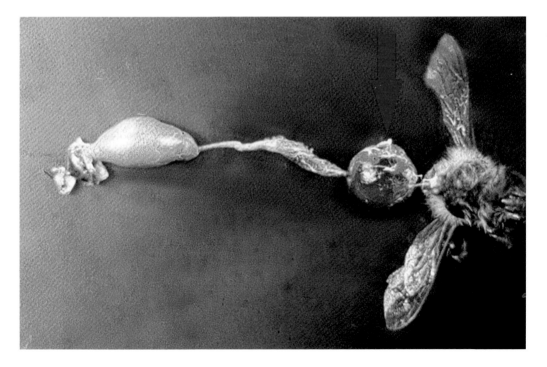

2. Also, CBPV may enter haemolymph through open wounds from broken hairs on adult bees. Hairs can be broken when foragers stay in the hive for too long, such as during poor weather.

CBPV causes two sets of symptoms (form 1 and form 2), both of which lead to the death of bees. They can occur simultaneously, but only one of them can be dominant in a bee, a genetically determined characteristic.

Form 1: The bees cannot fly and often crawl on the ground and up plant stems, with strange trembling of the body and wings. They may have a bloated abdomen due to an overextended honey sac. The wings are partially spread or dislocated. In a badly infected hive, 1000 or more bees can crawl on the ground. Infected bees huddle together on the top of the cluster or the top bars of the hive.

Form 2: Infected bees can fly, are almost hairless, look black, have a relatively broad abdomen, and otherwise look smaller. The older bees in the colony often bite them, and this behaviour may be the cause of their hairlessness. Infected bees are hindered at the entrance to the hive by the guard bees. A few days after infection, trembling begins. The infected bees then become flightless and will soon die.

Slow bee paralysis virus (SBPV)

SBPV was discovered in England in 1974 and infects honey bees through *Varroa* infestations. The virus accumulates mainly in the head, salivary glands and fatty tissues of a bee, and to much lesser extent in the hindlegs, midgut and rectum. Because of this, the virus may be spread by oral transmission between bees, during prophylaxis, feeding larvae, or collecting nectar from returning foragers. The virus causes paralysis in the front two pairs of legs of adult bees a few days before dying, which occurs approximately ten days after infection by the virus. The colony may also die if many bees are infected.

Larvae and pupae infected with SBPV do not show symptoms. Because

Figure 7.7: (a) & (b) Black hairless 'robber' bees, infected with CBPV, often lose hair and have a shiny black body.

SBPV is highly virulent, killing the host bee rapidly before significant replication, it is relatively rare in colonies, even those that have a strong *Varroa* presence.

The virus has not been found in the United States, although there is a low natural prevalence across parts of Europe.

Lake Sinai viruses (LSV)

Lake Sinai virus, of which there are many variants, has been detected in honey bees, mites and pollen. The virus only replicates in *A. mellifera* and mason bees (*Osmia cornuta*) and not in *Varroa* mites. Although LSV was first detected in *A. mellifera*, it has also been detected in solitary bees and bumblebees. The effect of LSV on *A. mellifera* is unknown, but infection has been linked to poor colony health and Colony Collapse Disorder (CCD).

Summary

Viruses cause infection by invading host cells, replicating within the cells, and eventually leaving the cell to infect other cells. As part of their survival strategies, bees have evolved mechanisms to defend against viral invaders by employing effective immune responses. We know this as viruses live in apparently healthy colonies without the bees showing any apparent sign of viral infection. Indeed, many of the viral infections detailed above become apparent and affect the health of the bee only when there is some other agent or another factor, such as stress or *Varroa*, to overload or compromise the bee's natural equilibrium.

CHAPTER 8

Pests

Although *Varroa* and *Tropilaelaps* may be
regarded as pests, this chapter focuses on
the less-harmful pests, wax moths and hive
beetles. Less-harmful, though, does not mean
harmless. Without adequate management
by the beekeeper, these pests may also cause
significant harm in the apiary.

Wax moth

Introduction

Wax moth can cause severe damage to frames of honey and brood, although infestation is usually limited to weaker hives that have insufficient numbers of bees to provide protective cover on frames to fight the moth.

There are two related, though physically different, types of wax moth:

- **the greater wax moth, *Galleria mellonella* (GWM)**
- **the lesser wax moth, *Achroia grisella* (LWM).**

The greater wax moth is the larger and more destructive of the two, with a female adult length of about 20 mm and a wing span of 30 mm to 41 mm (Figure 8.2a). She is a brown-grey colour and may have several speckles on her wings. The lesser wax moth has a female adult body length of about 10 mm to 15 mm and is a silvery grey colour (Figure 8.2b). The length of greater wax moth larvae when fully grown are up to 28 mm (Figure 8.1a), while lesser wax moth larvae are up to 13 mm (Figure 8.1c).

Greater wax moth

The female wax moth enters colonies during the night when the hive's defences are at their lowest (Figure 8.2). Once inside, the moth lays her eggs directly on comb or in other cracks that cannot easily be cleaned by the bees or seen by the beekeeper. Upon hatching, the larvae initially feed upon baseboard debris before burrowing into pollen storage or honey cells. As the greater wax moth larvae develop, they burrow towards the centre of the colony where they are protected by the capped cells, leaving behind a trail of destruction in the form of a dense mass of silken webs and faecal

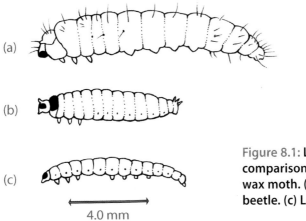

(a)

(b)

(c)

Figure 8.1: Larval size comparison. (a) Mature greater wax moth. (b) Small hive beetle. (c) Lesser wax moth.

4.0 mm

debris (Figures 8.3 and 8.4). Wax moth can destroy unattended comb in as little as ten to fifteen days, and the larvae will also devour young brood if there is insufficient other food available. Between 50 to 150 eggs are laid in each batch by a single female, and the maturation period for wax moth larvae, depending on temperature and food availability, is typically eighteen to twenty days in tropical regions and a little longer in temperate zones. In very cold climates the eggs can remain semi-dormant for up to six months until warmer conditions arrive. The newly emerged larvae are a creamy white colour but as they reach maturity turn grey. Infestation by the greater wax moth *G. mellonella*, often called galleriasis, can result in the young brood being unable to leave their cells due to entanglement in the silken threads of the wax moth cocoons.

Studies have shown that the newly hatched larvae of the greater wax moth can travel more than 50 metres (55 yards) and are therefore capable of moving to neighbouring hives. The average colony copes well and will control a small number of larvae, but if bee numbers are reduced, wax moth larvae may take over and destroy the hive.

Lesser wax moth

As its name implies, the lesser wax moth is smaller than the greater wax moth. When resting, both the greater and the lesser wax moths fold their

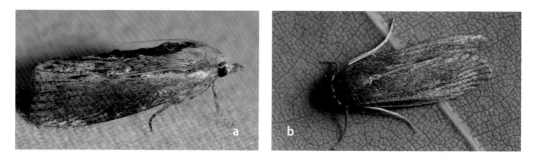

Figure 8.2: (a) Greater wax moth. (b) Lesser wax moth.

Figure 8.3: (a), (b) and (c) Damage caused by greater wax moth. Damage includes a silken mat that bees get trapped in and collapsed comb that the larvae have eaten.

wings over their body much like a cloak — although the lesser wax moth has more elliptical wings and the larger one has a hipped appearance like a roof of a house. The lesser wax moth has a distinctive yellow head and can often be found walking around the hive entrance looking for an opportunity to enter. Also, while the greater wax moth has a fatter body compared to its length, the lesser wax moth has a distinctive slender body and is silver-grey. The lifespan of an adult female is about seven days, during which time she lays around 250 to 300 eggs.

As with the greater wax moth, infestation by the lesser wax moth usually occurs in weaker hives, with the moth's larvae preferring to feed on dark comb containing pollen or brood. The larvae are generally white with a brown head and are often found in debris on the hive floor, although they do not congregate in the large wriggling grub numbers of greater wax moth larvae.

Careful identification of larvae by the beekeeper is necessary because wax moth and small hive beetle larvae are very similar in appearance. Wax moth larvae have three pairs of legs on the front of the body and have other uniform pairs of prolegs along the rest of the body. On the other hand, small hive beetle larvae have only three pairs of legs at the front of the body, and no prolegs are present along the rest of the body. Wax moth larvae are soft and fleshy, whereas the small hive beetle larval body is rigid and stiff. It is standard for both pests to be present in the same hive. If larvae are found in the hive, a simple way for the beekeeper to determine if a wax moth is present rather than a hive beetle is the presence of a mess of webbing in and on the comb. Webbing is a clear indicator of wax moth as small hive beetles do not create either webbing or the characteristic corrugated burrowing damage in the wooden parts of the hive or frame of the greater wax moth.

Spread

Since the primary food source for wax moths is found inside a hive or feral nest, the moth will fly from one colony to another in search of a new home or food.

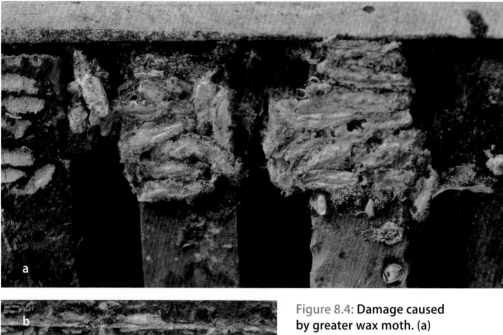

Figure 8.4: Damage caused by greater wax moth. (a) Frames partly eaten by greater wax moth, plus silken cocoons. (b) Inside of hive partly eaten by moth.

Signs of infestation

For both types of wax moth, the signs are the same: larvae crawling on the surface of the comb, frames and box sides. Larvae burrowing through comb, particularly brood comb, are often the first signs of infestation in hives. Cocoons can sometimes be found on the top of frames or in other hive parts. Cocoons are distinctive due to their larger size and are often covered in tiny black pellets of larval faeces.

Later stages of infestation are very distinctive due to the tangled mass of silk webbing that covers the parts of the frame where larvae are burrowing. When the infestation is even further advanced, the comb in the frame falls apart, a very dense mass of impenetrable silk matting has been spun, and larvae can be seen crawling over what is left of the comb.

Severely infested hives can result in significant damage to the wooden boxes and frames as the larvae chew the wood. Physical damage to hive boxes and frames appears as gouges or holes in the wood. Often wax moth larvae can be seen crawling out of these holes when disturbed.

The adult wax moth is harmless and does not damage hives; only the larvae can cause severe problems for the beekeeper.

Treatment

There are no approved chemical treatments for either the greater or lesser wax moth, and thus effective control comes down to management practices by the beekeeper. Since the moths target weaker colonies, the beekeeper should ensure that the hive remains strong by merging weak colonies. A strong hive is a healthy hive so ensure that plenty of pollen or pollen substitutes are available for the bees and that there is a suitable reserve of capped honey or syrup during periods of drought. If the hive has too many boxes so that the distribution of bees across the hive is sparse, this will allow the moth to flourish. It is, therefore, advisable to keep empty space within the hive to a minimum by limiting the number of supers added. A healthy hive will cope well with a small infestation of wax moth larvae, and the adult bees will quickly remove these. Like many pests and diseases affecting the honey bee, a weak colony will be most affected by wax moth larvae build-up and may in time be annihilated or abscond.

An effective method of treating empty supers and frames for wax moth infestation is to fumigate them inside an airtight container or bag using Phostoxin. The chemical name for Phostoxin is aluminium phosphide, which breaks down in the presence of water into the gas phosphine and aluminium oxide.

Wax moth also causes problems with frames containing used comb, unclean wax, or used hive boxes stored in the open or in sheds. To minimize infestation, frames should be stored in airtight plastic containers, preferably in cool rooms. If the beekeeper plans to keep capped or partly capped frames of honey for any period, any wax moth eggs or larvae can be killed by wrapping the frames individually in sealed plastic bags and freezing for two days. After that time the frames may be safely stored in airtight plastic boxes as all the larvae will have been killed. If there are too many extracted frames to return to populated hives, return the frames containing pollen first since pollen is attractive to the moth, and this will likely result in less infestation by the moth. Frames containing pollen are far more appealing to the moth than frames containing only dark comb. Keeping the extracted frames that do not have pollen in the apiary shed or cool room is less likely to be attractive to the wax. This procedure does not affect the taste or quality of the honey in the frame.

Wax moth will settle and breed only in dark locations, and knowledge of this behaviour can be used to the beekeeper's advantage. Empty frames should be stored inside supers and stacked after placing sticks between each box to let in light. A lid is unnecessary, since as much light as possible should be allowed to enter through the top of the stacked boxes.

Small hive beetle (SHB)

Introduction

Small hive beetle, *Aethina tumida*, was previously known only from the sub-Saharan regions of Africa where it has been considered a minor pest of bees (Figure 8.5). While honey bees in sub-Saharan Africa have adapted to small hive beetles and can manage the pest well, SHB can be a major pest of the European honey bee, causing significant economic damage.

The small hive beetle mainly affects weaker hives as there are insufficient worker bees to remove or drive the beetles from the hive. The adult beetle lives in areas of the hive box inaccessible to bees, such as under the hive mat, or in places less frequented by bees, like the corners of lids and cracks in the wood of the box.

As shown in Figures 8.1 and 8.6, the larvae of small hive beetle are often mistaken for wax moth larvae, although there are significant differences between the two:

- **Small hive beetle larvae are smaller than wax moth larvae. SHB larvae are around 11.1 mm long and 1.6 mm wide when fully grown. Greater wax moth, *G. mellonella*, larvae are up to 28 mm (1 inch) long, and the lesser wax moth (*A. grisella*) larvae up to 13 mm (1/2 inch) in length.**
- **The SHB larva has three pairs of legs at the front of its body.**
- **The SHB larva has a row of spines along its back together with two distinct spines protruding from the rear.**

Adult female small hive beetles will lay eggs directly onto food sources within the hive, such as pollen or brood combs. She may also deposit irregular masses of eggs in tiny crevices or cavities away from the bees. A female small hive beetle may lay 1000 eggs in her lifetime, although some researchers suggest that the number of eggs produced in one female's lifetime might be upwards of 2000. The majority of these eggs hatch within three days.

Newly hatched larvae immediately begin feeding on the available food including honey, pollen and bee brood. Maturation time for larvae is generally ten to sixteen days. Once the larvae finish feeding, a wandering phase starts where the larvae leave the hive to find suitable soil in which to pupate. It is believed that most larvae do this at night under the cover of darkness.

Larvae in the wandering stage may travel considerable distances from the hive to find suitable soil. Despite this, most larvae pupate within 90 cm (35 inches) of the hive. Larvae will burrow 10 cm (4 inches) into the soil to pupate. The time spent in the ground pupating can vary greatly depending on soil temperature, moisture and soil composition factors. However, most adult beetles emerge from the soil in approximately three to four weeks.

Upon emerging from the ground as adults, the small hive beetle flies in search of honey bee colonies and probably identifies potential host colonies by a range of smells emanating from the hive, including honey and pollen.

The beetle flies before or just after dark and, upon locating and entering the host colony, will seek out cracks and crevices where she can hide from the bees' aggression.

The life cycle begins when the adult enters the hive and lays her eggs.

The turnover rate from egg to adult emerging from the soil can be as little as four to six weeks; consequently, there may be as many as six generations in twelve months under moderate climatic conditions.

Small hive beetles are a problem in hot, damp climates. Even though there may not be many small hive beetles in the drier areas, if discarded comb or residue from cleaning hives is left on the ground, it may become sufficiently damp to host the beetle.

Small hive beetles can host several honey bee viruses, *Nosema ceranae*, and other gut parasites. The extent to which SHBs serve as reservoirs and spread these diseases, as well as whether these diseases adversely affect small hive beetles, is poorly understood.

Spread

The beetle is attracted to the smell of honey and flies from one hive to another. The migratory nature of commercial beekeeping means that the beetle can easily be moved around, infecting new hives wherever it goes. It is believed that the beetle can fly up to 10 km (6 miles), and thus a single infested hive can infect other colonies over an area of over 300 sq. km (115 square miles). Since feral colonies also harbour the beetle, the movement of commercial hives into new regions can inadvertently lead to infestation from feral colonies.

The small hive beetle, although not yet widespread globally, is slowly spreading worldwide. SHB has been reported in southern Africa, Europe, North America, Mexico, Bangladesh, parts of Central America and Australasia. The widely distributed, sporadic areas to which SHB has spread strongly indicate they have been taken to these places, unintentionally, by human activity.

The beetle can also live in the nests of solitary bees, making them a threat to many species of bees.

Figure 8.5: (a) Adult SHB inside a hive. (b) SHB larvae that have slimed the honey frame. (c) SHB eggs. (d) SHBs can eat pollen on flowers and are not reliant on honey bee colonies for their survival.

Figure 8.6: Comparison between small hive beetle and greater wax moth larva, not to scale. SHB larvae do not spin a cocoon as do wax moth larvae. Beetle larvae have tough exterior bodies; wax moth larvae have soft exterior bodies that can be penetrated easily. If you can easily squish it between thumb and forefinger, it's a wax moth; if not, it's a hive beetle. (a) Small hive beetle, *Aethina tumida*, larva has three legs at front, no prolegs in centre, a ridge of spines on top, and two prominent spines at the rear. (b) Greater wax moth, *Galleria mellonella*, larva has three pairs of legs at front, four more prominent pairs of prolegs at centre, and no spines on top. Fully grown greater wax moth larvae are larger than mature SHB larvae.

Signs of infestation

Depending on how frequently hives are inspected, the first sign of small hive beetle may be:

- Opening the hive and finding the beetles scurrying away out of sight, say from on top of the hive mat.
- For the less frequently inspected hive, noticing a slime oozing out of the hive entrance and a generally putrid smell similar to that of rotting oranges. The small hive beetle is not the cause of the destruction; rather it is the larvae. As the

larva grows, it burrows its way through the comb, preferably through a comb containing brood or pollen. During burrowing, it leaves a residue of yeast in its excreta that quickly causes the contents of the affected comb to ferment, slime and fall to the hive's base as a revolting slurry.

Either of these is a sign of an infestation of small hive beetle that needs managing. Other signs to look for and ways to check for an infestation:

- Observe whether small larvae are burrowing through capped comb, both honeycomb and brood comb. Since the comb is capped, it may be difficult for the beekeeper to notice the larvae during a routine hive inspection.
- Look for beetles inside the non-capped comb. Tap the frame onto the lid and observe if any beetles fall out.
- Inspect under the lid of the hive, on top and underneath the hive to see if there are any beetles there.
- Pick up the bottom brood box and check if there are any larvae or beetles on the baseboard, particularly if the baseboard is dirty and the beetles may be hiding in the debris.
- If the brood box is attached to the baseboard, remove all frames from the brood box and inspect the baseboard from the top of the hive.
- Check all the nooks and crannies in the hive where beetles may be hiding.
- If plastic frames are used, check that beetles are not hiding in the hollow ribs of the frame's sides — check metal queen excluders with a strip edge as these can also provide a suitable hiding place.
- Remove a super with frames and place it upside down on a lid. Wait for several minutes before returning the super to the hive and quickly inspect the lid for any beetles that may have assembled to hide from the light.

Treatment

As previously noted, strong hives are reported to be less susceptible to small hive beetle than weaker hives, although this has been disputed by some experts, particularly in areas where the beetle has gained a good foothold. To obtain stronger colonies, merge weaker colonies or add a frame or two of capped brood to the weaker hive. Either of these options ensures that you are not transferring any disease from an infested colony to an uninfested one.

Small hive beetles are attracted to the smell of honey and use it to guide them to hives. Consequently, to reduce the chance of the beetle being attracted to your hive by the smell, minimize the number of times you open the hive and keep the area around your hive clear of old honeycomb and debris that may act as an attractant. Small hive beetle also flies with swarms, so by catching a swarm you may unwittingly introduce the beetle into your apiary.

The beetles prefer to breed in a humid environment. Keep the inside of hives a little drier; open the lid by a minimal amount for a few hours to allow any moisture to escape, say by placing a twig under one side. The use of a vented base may also keep beetle numbers down. Leaving the hive like this for longer periods will encourage robber bees, wasps or even more beetles to enter, so this option can be time-consuming and is really for the hobby beekeeper with a small number of hives.

The beekeeper should keep the hive boxes in good condition, repairing any cracks that the beetle can use to enter the hive and hide within. Also, ensure good hygienic practices in and around the hive by keeping the area around the hive clean and tidy, clearing any burr comb from inside the hive lid, and removing waste material from the hive floor. This last practice is made a lot easier if baseboards are not fixed to brood or bottom boxes.

Minimizing the movement of the hive is also a good idea since by changing its location, the beekeeper may be moving it close to a colony that already has the beetle and will unwittingly aid in its spread.

Apart from the above-mentioned general management practices, the following techniques have been successfully used to control the beetle:

Figure 8.7: **Small hive beetle traps. (a) Reusable trap. (b) Reusable trap between frames in hive.**

- Beetle traps can be placed inside the hive between frames. There are several of these types of beetle traps for sale. The bees chase the beetles into the trap where they drop into the reservoir and are killed either by drowning in a mixture of vegetable oil and a touch of apple cider vinegar or smothered by garden lime or diatomaceous earth (Figure 8.7). However, the use of diatomaceous earth in hives is not approved in all countries. Before using it, check if it can be used in your country. Any kitchen vegetable oil can be placed in a trap. This option has the advantage that the oil mixture is non-toxic, does not have an unpleasant taste and will not contaminate honey if accidentally spilled onto the frame. Some traps use a clear plastic reservoir, which makes it easy to carry out a dead beetle count and check the level of oil.

- Another popular method used to trap beetles is carpet underlay or felt-backed linoleum as a hive mat. The fibrous mat traps the beetles, which can be disposed of during the next inspection of the hive by the beekeeper. A minor disadvantage of this method is that the odd bee will often be caught in the felt. Disposable kitchen wipes and some types of carpet have also been shown to be effective at trapping SHB. Success can often depend on the beetle numbers within the particular hive and how vigorously the bees pursue them.
- Corflute, made of cardboard or plastic with a hollow, tubular core, can be inserted into the hive entrance or placed under the lid. The beetles will use the hollow tubes of the Corflute to hide from the bees, and the Corflute can then be removed by the beekeeper and replaced. Ensure that the Corflute can be conveniently inserted and removed from the hive entrance; attach a piece of stiff wire to it so that it can be pushed in or pulled out easily. See instructions on page 135 for making a small hive beetle trap out of Corflute.
- An effective variation of Corflute, the Apithor Hive Beetle Harbourage has a hollow fluted matrix impregnated with the miticide fipronil. The Apithor trap is placed through the hive entrance and onto the baseboard of the hive. The beetles enter the hollow matrix to escape the bees and are killed by the miticide in the trap. The miticide is only on the inside of the matrix where the mites hide so cannot be touched by the bees. This device is effective in controlling beetles, and the matrix is small enough to be easily inserted and removed from the baseboard of the hive. (The use of Apithor within the hive is for surveillance purposes only and is currently not legally approved as a treatment for SHB within Australia.)
- Another method of control is to use a specially constructed base with a tray of vegetable oil, garden lime, or diatomaceous earth with wire gauze above the tray so that the beetles

can climb through the wire but the perforations are too small for the bees to enter. The beetles fall into the tray where they are killed and can be cleaned out by the beekeeper. For ease of use, the tray must be removed from the base without disturbing the rest of the hive. If the hive needs to be dismantled to remove the tray of dead beetles, this will mitigate many of the technique's advantages.

- The design of traps for the beetle is limited only by imagination, and beekeepers have successfully used empty CD/DVD cases, fishing tackle boxes, and containers previously used for screws as inexpensive but effective beetle traps.
- The following two techniques to control the beetle's spread are better suited to professional apiarists. Since the pupation of larvae, an important part of the beetle's life cycle, takes place in the ground up to 100 metres (110 yards) or more from the hive, the chemical permethrin can be used to drench the ground to kill the pupating larvae. Since permethrin is dangerous to bees, care must be taken to ensure that none of the permethrin comes in contact with the foraging bees or guard bees at the entrance to the hive. This is best achieved by spraying the area around the hive at night or a few days ahead at the new location before the hive is moved there. To minimize permethrin drifting towards the hive as a fine spray, use a watering can with larger holes to spray the chemical, as droplets of this size are unlikely to drift.

Chemical treatment

'Trap and kill' methods are most frequently used by small-scale beekeepers; although this method is environmentally friendly, it is impractical for large-scale apiarists. Most commercial beekeepers use some sort of miticide to control the beetle. In the United States, only one insecticide has been approved for in-hive use against hive beetle larvae: coumaphos, an organophosphate sold as CheckMite+ strips. The recommended use for

CheckMite+ is to cut a strip in half and staple each piece to the centre of the corrugated side of a 15-cm (6-inch) square of corrugated plastic or cardboard. The corrugated square is then placed on a solid bottom board with the strip facing down where bees cannot reach it. The strip should be attached firmly to the baseboard so the bees cannot drag it out of the hive. Many beekeepers are concerned about this method since coumaphos is readily absorbed into comb wax. Coumaphos can accumulate in the wax, potentially poisoning adult bees and brood. Due to its toxicity, it is recommended that you replace your old comb with a new foundation every three years to minimize build-up in the hive.

A permethrin-based liquid product like GardStar 40 or Permethrin SFR is often applied to the soil around hives that have been slimed out or killed by SHB. This will kill wandering larvae and pupae, but this treatment should not be used if there is no infestation.

To minimize the contamination of stored frames with the beetle, if possible keep them in a cool room below 10°C (50°F) to be effective.

Make a simple small hive beetle trap

There are several inexpensive homemade traps for SHB, including drilling holes in a CD/DVD holder containing a small amount of oil, or placing a sheet of floor linoleum with a rough fibre backing that traps the beetles. These traps are good for hobby beekeepers but less practical for larger-scale beekeepers. A popular trap that is inexpensive, easy to make and can be reused is made out of Corflute (Figures 8.8, 8.9 and 8.10).

These traps can be stored for a few days in a plastic container before placing in the hive. Several can be placed in a heavily infested hive to manage the mite.

After a few weeks, either when the trap is blocked with beetles or the attractive smell of cockroach poison has disappeared, remove the trap, remove the tape, bend it over to expose the inside ribs, place it in a mesh bag, and wash in a washing machine. The clean trap can now be stored and reused.

Figure 8.8: **Cut a square of Corflute 6.5 cm x 6.5 cm (2.5 inches x 2.5 inches). Make a cut on the top of one side across the Corflute ribs. Bend the Corflute, exposing the tubes that make up the inside.**

Figure 8.9: **Using a cockroach miticide applicator, place miticide along the cut inside the Corflute. Fipronil-based cockroach poison is frequently used in the trap.**

Figure 8.10: **Flatten the Corflute, and use duct tape, electrical tape or similar to seal the trap and hold it together.**

Large hive beetle (LHB)

Introduction

The large African hive beetles, *Oplostomus fuligineus* and *Oplostomus haroldi*, are pests of honey bee colonies in southern and central Africa. Adult beetles feed on brood and pollen, damaging the combs (Figure 8.11). A single honey bee colony can have over 700 adult beetles present. Unlike small African hive beetles, *Aethina tumida*, the large African hive beetle does not develop within colonies. Instead, it lays its eggs in horse dung but may also lay in the dung of cattle.

Large hive beetle prefers drier grazing land where dung is readily available, but it is absent in dry areas like the Kalahari Desert in Botswana. Large hive beetle causes significant problems to beekeepers. Due to its limited distribution in Senegal, Nigeria, Kenya, Zambia, Namibia, Botswana and South Africa, it is not a serious threat to bees globally.

Large hive beetle female and male adults can be distinguished by the shape of the abdominal plates on the upper sides of the body. They prefer to feed on open bee brood containing larvae and young capped brood and survive well on pollen and honey. In nature, they seldom feed on flowers.

Figure 8.11: **Large hive beetles. (a) Adult *Oplostomus fuligineus* feeding on brood. Notice the black head.**

Figure 8.11: (b) *Oplostomus haroldi*, photo taken in Botswana. Notice the red head. (c) *Oplostomus fuligineus* larvae. Newly formed larvae are creamy white but become a light reddish-brown colour before becoming an adult. (d) LHB trying to enter a hive.

Braula fly

Introduction

The *Braula* fly, *Braula coeca*, is often incorrectly called the bee louse. The fly is found in many parts of the world (Figure 8.12). Adult *Braula* flies are reddish brown and are regularly misdiagnosed as *Varroa* mites. One of the main differences is that *Braula* has six legs, while *Varroa* has eight legs (Figure 8.13).

The *Braula* fly is unusual in that it is wingless, flattened, and lives in honey bee colonies by holding on to the body hairs of adult bees, usually on the head, using a set of comb-like structures on its front legs. When hungry, the fly moves to the mouth of the bee and steals some of the food. There is some evidence that *Braula* can induce a single bee to regurgitate honey by stroking the bee's labrum, roughly corresponding to a bee's lip.

The *Braula* fly does not damage or parasitise any stage of the honey bee

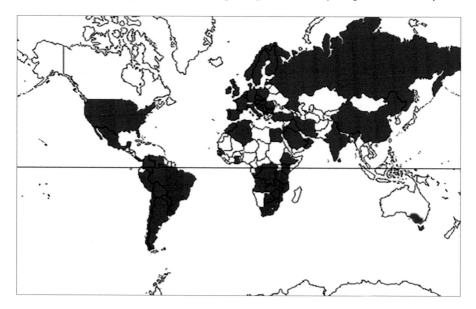

Figure 8.12: **Global distribution of the *Braula* fly.**
Equator shown as a line at centre of map.

Figure 8.13: (a) *Braula* fly on a worker bee. (b) Dorsal view of a *Braula* fly. (c) Side view of a *Braula* fly. (d) Queen infested with multiple *Braula*.

life cycle and is not considered a significant pest or threat to colonies of honey bees. If *Braula* is living on a queen, it may reduce the amount of food available to her, resulting in an impaired ability to lay eggs.

The main economic impact of the *Braula* fly occurs due to the fly laying eggs on capped honey and the larval stage burrowing under the honey cappings. The sight of *Braula* larva burrows in comb honey can detract from honeycomb intended for sale. Liquid honey is not affected by *Braula* since most honey is extracted mechanically and will be filtered as part of the extraction process.

Spread

The *Braula* fly can spread by being attached to bees during swarming, absconding, robbing, or attached to bees drifting to other colonies. *Braula* fly can also spread through the interchange of hive components from apiary to apiary and by the movement of hives.

Signs of infestation

Observation of adult bees, particularly the queen bee, can reveal the presence of adult *Braula* flies. Close examination of bees in the hive using a magnifying glass effectively identifies *Braula* on their hosts.

Braula is found on the head of honey bee workers, drones and queens. As a result of this preference, queen bees should be thoroughly and regularly checked.

The fly can be confused with the parasitic *Varroa* mites (*V. destructor* and *V. jacobsoni*) and Asian mites (*Tropilaelaps clareae* and *T. mercedesae*). As an aid to identification:

- **Adult *Braula* flies are less than 1.5 mm long and covered in spine-like hairs. They do not have wings as most flies do.**
- **A notable difference between *Braula* (an insect) and *Varroa* (a mite) that is useful as a field diagnostic tool to distinguish**

between the two is the presence of six legs on the *Braula*. In comparison, adult *Varroa* mites have eight legs. Further, the adult *Braula* fly has a rounded appearance, while *Varroa* mites are more compressed and oval. Despite these differences, both are very small and difficult to distinguish with the unaided eye, and inspection is best made using a magnifying glass.

- Adult female *Varroa* mites are oval, flat, reddish brown, and are 1 mm long and 1.5 mm wide.
- *Tropilaelaps* mites are active, red-brown mites that are around 1 mm long and 0.5–1 mm wide.
- *Braula* flies are elongated and about 1.5 mm long with eight legs, and they are generally attached to a bee's head.

The practice of uncapping honeycombs as part of extraction is an effective means of controlling the larval stage of the fly. To further prevent the larval stage, comb honey should be stored in a freezer for 48 hours as soon as it is removed from a beehive. This will ensure that all stages of the *Braula* brood life cycle are killed and will also kill other beehive pests such as wax moths and small hive beetle.

Treatment

Recommended treatments have not been established since the fly is not regarded as a destructive pest to bee colonies. Applying good management practices is the best method of controlling this pest by maintaining healthy, strong and vigorous colonies that display good hygienic traits.

Melittiphis

Little is known about the pollen mite *Melittiphis alvearius*, which has been found in Europe, the United Kingdom, New Zealand, Australia, Korea,

Canada, the United States and South America. Its life cycle is largely unknown because, unlike other mites that infest colonies, *Melittiphis* does not harm bees. Instead, it feeds on stored pollen and only uses adult bees for dispersal (Figure 4.1). Because of this, *Melittiphis* is not considered a parasitic mite. Beekeepers must be able to recognize this mite since it lives in colonies and may be mistaken for *Varroa* or *Tropilaelaps*.

There are many other insects that live in hives that are not believed to be harmful to the colony. The wax beetle, *Platybolium alvearium*, found in many parts of Asia, is an example.

CHAPTER 9

Other problems

Chilled brood

A healthy colony containing a large number of workers can keep its brood at 34°C to 35°C (93 to 95°F), which is the optimum temperature for raising brood. This temperature needs to be regulated if the colony is to remain healthy and produce a viable brood. If for some reason regulation fails, the developing brood may become chilled to the point where development stops and death occurs.

There are several causes of chilled brood, including:

- **Smaller hives or nucs that have an insufficient number of workers to keep the colony warm.**
- **Hives that are rapidly expanding during spring that contain large numbers of brood but do not yet have sufficient numbers of adult workers to keep the brood warm.**
- **Hives that have lost large numbers of adult bees due to disease or pesticide poisoning.**
- **Opening hives in cold weather, allowing warm air to escape.**

The best way to minimize chilled brood is to keep strong colonies and not to open hives in cold weather. Unless under exceptional circumstances, hives should not be opened if the temperature is below 18°C (64°F), higher if there is wind chill. Another technique is to reduce the entrance size during the winter to minimize the amount of cold air flowing through the colony. In extreme cases, it is necessary to have an entrance at the top of the hive just under the lid to prevent snow build-up close to the hive entrance. In some countries, such as Canada or parts of Europe, where winter temperatures can remain well below freezing for long periods, other actions may be necessary, such as insulating hives or keeping hives indoors.

Overheating of the hive

Another problem is hives overheating due to hot weather. Daytime temperatures in some parts of the world, such as Australia or parts of the southern United States, may reach over 40°C (104°F) during summer, and this can prove detrimental, stressful and at times fatal to the colony. To minimize the effect of heat, the following practices have proved to be effective:

- Where possible, locate hives so that they are in the shade during the hottest part of the day, usually between 10 am and 6 pm.
- Keep the entrance to the hive as open as possible.
- Do not open the hive when the air temperature is above 33°C (91°F).
- Ensure air vents at the hive lid are adequate and are unblocked to enable the fanning bees to keep cool air flowing through the hive.
- Try using seven frames in an eight-frame super or brood box, or nine frames in a ten-frame box, to allow for greater air circulation around the frames.
- Keep hives near water as this will cool the air and provide drinking water for the colony. For the hobbyist with a few hives, there are additional possibilities. The top of hives can be watered, taking care not to drown any bees before it becomes too hot. This and water under and around the boxes frequently during extreme weather conditions will assist the bee's efforts inside the hive where collected water is fanned with numerous wings to produce an air-conditioned cooling effect. Other hobbyist beekeepers place old beach umbrellas over their hives and water the top of them to offer some relief to bees in the hives below.
- When transporting hives, ensure good airflow through the hive and that the entrance is not blocked off. Perforated hive

entrance closers can prevent bees from escaping while allowing air to pass in and out of the hive. The use of a screened
travelling top is recommended as, again, it allows air to cir-
culate and provides the bees with fresh air.

- Paint hives with white or other light-coloured paint as darker
  colours attract and hold heat.
- Provide insulating shade covers on top of the normal hive
  lid, particularly if this is metal.

Since fire is a major concern to beekeepers, keep hives away from long
grass or vegetation that can burn. Reduce vegetation to a minimum around
hives, particularly during hot and dry periods.

Bees are affected by the heat, and they can become very aggressive and
challenging for the beekeeper to manage without suitable protective clothing and copious amounts of smoke. It is not a good idea to work hives in
hot weather as this destroys the interior temperature the bees have fought
so hard to maintain and will take a considerable time for them to re-reg-
ulate. The best option is to leave hives alone until the temperature drops,
hopefully after a few days. If a hot day is forecast and it is necessary to open
the hive, this should be done in the early morning when it is cooler or, if
required, in the late evening when the bees will be inside. Always be aware
that a total fire ban may be in place on sweltering days. Unless an alternative liquid calmer is available any hive inspections will need to be done
without the benefit of a smoker.

Damp hives

The classic sign of a damp hive is condensation inside the lid. The condensation will pool into droplets, which will drop down upon the bees in the
hive. Dampness in hives contributes to or causes many health problems in
bees, including *Nosema* and sacbrood. If dampness in hives is a problem,
try the following:

- Add ventilation at the top of the hive either by providing air vents in the lid or by drilling 2.5 cm/1-inch-diameter holes at the top of the uppermost super. Holes may be covered with mesh to stop robber bees. Another technique is to place lollipop sticks, twigs or small stones under the lid rim, raising the lid by about 3 mm, thus providing greater airflow throughout the hive. Any larger gap may open the hive to predators or robbers.
- The base of the hive needs to slope forward, allowing condensation to flow out through the entrance and not pool at the back of the base.
- Place the hive away from the damp ground and preferably raise it off the ground by about 0.5 metres (1.5 feet).
- Place the hive in a location that is sunny but not too hot.

Scattered or spotty brood

A scattered or spotty brood pattern often signifies a failing queen or disease in the hive (Figure 9.1). If the queen is older, her collected sperm may be nearing exhaustion, and she may be unable to lay a fertilized egg in every worker cell. If this is the case, check for supersedure cells. If they are present, the workers have noticed her poor performance and are preparing to replace her.

Another cause of spotted brood is a disease such as EFB, AFB, sacbrood, chalkbrood, or a virus. Open the hive and inspect every brood frame for signs of illness. If there is any suspicion of AFB, obtain a specimen for analysis and, if confirmed, depending on local regulations, either start treatment or destroy the colony and hive. (See Chapter 5: Brood Diseases, for details.) If, however, the cause of the brood pattern is either a failing queen or other suspected disease, requeening is worth a try to rectify the problem. Inbreeding in closed populations may also cause spotted brood patterns.

Figure 9.1: Scattered or spotty brood. (a) Good brood pattern. (b) Poor brood pattern.

Drone-laying queen

A healthy queen will lay a fertilized worker egg in every worker cell. If the queen is older and running out of sperm, or if she has a defect in her oviduct that does not allow sperm to fertilize the eggs, she will lay unfertilized drone eggs in worker cells. Those cells will have the characteristic bullet-shaped drone cap (Figure 9.2). In this instance, the queen is referred to

Figure 9.2: Drone laying queen or laying workers? There is a mix of workers' brood and drone brood in the worker cells. Drone brood should be in larger drone cells. There is also only one larva in each cell; a laying worker would deposit multiple eggs in each cell.

as 'a drone layer'. If on inspection the hive contains large numbers of brood and adult bees, you can try requeening. Conversely, if the drone-laying hive has only a small number of bees left, it can be merged with a stronger hive after first killing the old drone-laying queen.

Multiple eggs per cell

If a colony has been without a queen for several weeks, the ovaries in some normally infertile workers change, and this allows some workers to lay unfertilized drone eggs (Figure 9.2). A typical sign of this is the presence of several eggs in each cell. If the queen has been absent from a hive for a

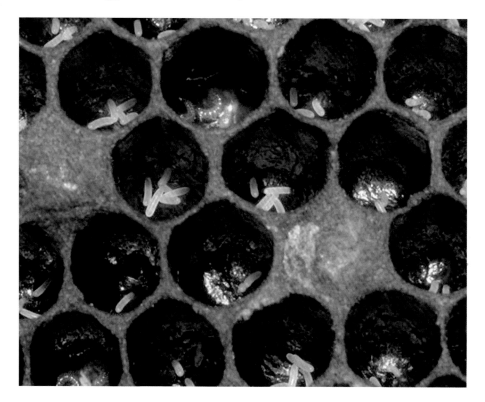

Figure 9.3: **Multiple eggs per cell.**

period sufficiently long for the workers to lay drones, requeening will not usually work. A course of action is to open the hive near other healthy hives, shake out all of the bees onto the ground then remove the hive. The bees will soon drift to healthy hives or will be lost.

Queenlessness

It is often very difficult to find a queen in a busy colony and the best way to determine if a laying queen is present is to look for cells containing either eggs or brood. Even this is not foolproof evidence, as a hive with a virgin queen may not have brood present, and it is best to wait for approximately two weeks to see if a fertilized queen starts laying eggs. If no eggs appear after two weeks, the hive is queenless and you will either need to introduce a new queen or merge the colony with another strong hive. For the observant beekeeper, a queenless hive is often apparent by the unsettled, confused, agitated nature of the bees both in and outside. Many also claim that these hives make a roaring noise when listened to closely.

If the colony remains queenless for several weeks, some workers will start laying multiple drone eggs in each cell. (See Multiple eggs per cell above.)

Shortage of honey or pollen

Shortages of honey or pollen may occur at any time during the year. However, the most common times are during the late autumn and winter and in the arid months of summer. Also, the beekeeper should be aware of the possibility of shortage during a rapid spring build-up of the colony when the foraging bees cannot keep up sufficient supplies to meet the demands of the developing brood.

If nectar is in short supply, the colony can be supplemented with sugar syrup, made using refined white sugar dissolved in warm water.

Alternatively, white sugar crystals can be placed either on top of the hive mat or in a frame feeder inside the brood chamber of the hive. Raw and brown sugar are unsuitable for bees and will upset their digestive tract, resulting in diarrhoea.

Pollen is an essential component of bee bread, and if the colony is to raise a healthy brood, this aspect of feeding must not be overlooked. It was previously common practice to feed colonies with a pollen patty (either brought commercially or made with a mix of sugar water and pollen) or a pollen substitute, such as yeast. With the advent of the small hive beetle, however, this practice has fallen out of favour in those areas affected by the beetle. SHB larvae are attracted to the protein in the patty and infest it quickly. Nowadays, most beekeepers feed dry powdered forms of pollen or pollen substitutes either by sprinkling them on the top bars of frames or by inserting them into the hive's base on a flat tray.

Colony Collapse Disorder (CCD)

It is important to realize that a collapsed colony and Colony Collapse Disorder are not the same thing. A collapsed colony is more specific and can be caused by a variety of often identifiable ailments that are affecting the bees in a colony. CCD is a very general disorder found in the United States, the causes of which are poorly understood, and it is not now regarded as being a major problem to American beekeepers. CCD can be defined by a condition that leads to the workers leaving the hive so that the queen is left with only a tiny number of workers. Often there is not a large number of dead bees in the vicinity of the hive, but it is as if the workers just left the queen behind and did not return to the hive after foraging.

Once CCD was first identified in the United States in November 2006, there was a concerted effort to identify a single cause for this large yearly loss of colonies, sometimes reaching as high as 30 to 45 per cent of all United States hives lost during the winter months. The United States Department of Agriculture, however, now acknowledges that CCD is not caused by a single pathogen but by a range of issues, some of which are due to human activity.

- Infestation by *Varroa*, the most harmful infestation of the honey bee, is closely associated with high winter losses of colonies.
- Multiple virus species have now been found in both bees and *Varroa* mites, including deformed wing virus, Israeli acute paralysis virus, and black queen cell virus, and are believed to be significant factors in poor colony health.
- *Varroa* is known to increase the levels of viruses in bees, acting as a vector for viruses.
- The bacterial disease European foulbrood is being detected more frequently in colonies and may be linked to colony loss.
- Nutrition has a major impact on colony health and longevity. The increased use of monocultures where a single crop is grown over very large areas and bees are transported there for pollination, such as for almonds, is proving harmful to bees due to the lack of a balanced diet that they offer.
- Gut microbes used by the honey bee to break down pollen, fight some gut diseases, digest food, transport nutrients from the gut into the body, and detoxify chemicals are harmed by the increasing use of chemicals for pesticides, miticides and herbicides, which the bees unintentionally digest.
- Transporting hives sometimes many thousands of kilometres each season for pollination is detrimental to bees' health.
- The use of high fructose corn syrup (HFCS), which is used widely in the United States as a food supplement, is not as effective as honey in developing and keeping healthy the microbes within the gut of the bee.
- The use by local councils and domestic homeowners in the United States of herbicides to eliminate 'unsightly' broad-leafed weeds on grasslands, parks and lawns is depriving

bees of both valuable food and further adding toxins to their bodies.

- Chemicals used within the hive to manage *Varroa*, small hive beetles and tracheal mite are proving more of a problem to bees than as a solution to pest control.
- Acute and sub-lethal effects of pesticides on bees have been increasingly documented and constitute a significant concern to beekeepers.
- In some situations, two chemicals applied to bees are significantly more harmful than either chemical used singly. These effects, together with the long-term effects of apparently harmless chemicals, are not well understood and need to be further researched before we can say with confidence what harm they are doing to all animals.
- Poor management practices by the beekeeper.

Many of the contributing factors listed above come from issues that are unrelated to each other. As a result, there will not be a single silver bullet that will eliminate CCD. The solutions to the problem will need to come from diverse areas. They will involve many groups of people working together, who even today could spend more time talking with each other before this cause of large-scale bee death is minimized.

Although CCD was, around 2006 to 2010, a significant cause for concern in the United States, it needs to be viewed within the broader context of other honey bee morbidities occurring worldwide. These morbidities can be explained by known pathogens or beekeeper management issues. One example is the harm caused by beekeepers' inability to control *Varroa*, which not only feeds on host fat bodies and weakens host immunity but also is a vector for a variety of viruses. It is now widely believed that honey bees' health problems are caused by multiple factors, both known and unknown, acting singly or in combination. Some of these factors are pathogens, and others are due to human activities, such as the excessive use of monocultures or the inappropriate use of agrochemicals. Many of the factors are outside of the control of beekeepers. The use of integrated pest management practices, both in agriculture and by beekeepers, will go some way to solving many of these problems.

CHAPTER 10

Other types of honey bees and hornets

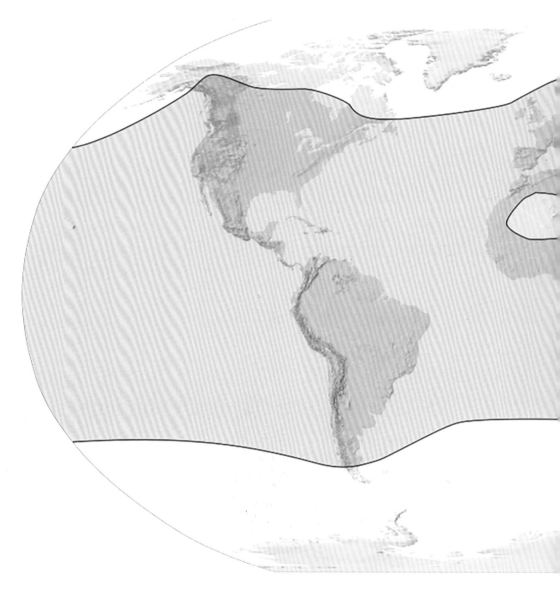

Figure 10.1: **Distribution of honey bees globally. The wide range of *A. mellifera* is due to it being moved for pollination and honey production to most parts of the world.**

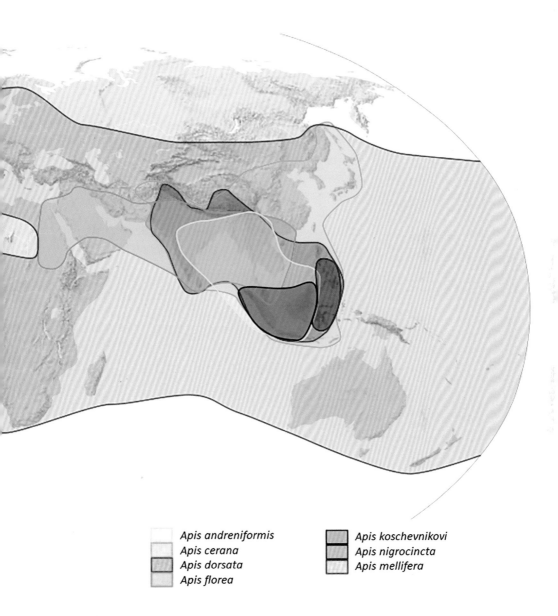

Apis andreniformis
Apis cerana
Apis dorsata
Apis florea

Apis koschevnikovi
Apis nigrocincta
Apis mellifera

Introduction

There is disagreement among scientists about the number of honey bee species, of the genus *Apis*. Currently, eight honey bee species are recognized globally, although, historically, seven to eleven species were recognized. Only *Apis mellifera* is native to Europe, Africa, the Middle East and a small part of central Asia (Figure 10.1). The remaining seven are native to Asia and are present in South East Asia. Honey bees are classified into three groups:

The giant single-comb open-air-nesting honey bees
1. *Apis laboriosa*
2. *Apis dorsata*

The medium-sized multiple-combs cavity-nesting honey bees
3. *Apis mellifera*
4. *Apis cerana*
5. *Apis koschevnikovi*
6. *Apis nigrocincta*

The dwarf honey bees, which build small single-comb nests in the open air
7. *Apis florea*
8. *Apis andreniformis*

The giant Asian honey bee, *Apis dorsata*, is *Tropilaelaps'* primary host, although it may also be found in the colonies of other Asian honey bees, including *A. cerana* and *A. florea*, as well as in *A. mellifera* colonies in Nepal. All these species are infested or infected with pathogens, some lethal while others are benign. Many of the pathogens that are lethal to *A. mellifera*, such as *Varroa* and *Nosema ceranae*, originated with other species of honey bees in Asia. The description of pathogens of non-*mellifera* bees, which have not jumped species to *A. mellifera*, is beyond the scope of this book and will not be discussed further.

The Western honey bee, *Apis mellifera*

Apis mellifera is often called the Western or European honey bee, although the two terms are not interchangeable. The Western honey bee refers to all *A. mellifera* subspecies or strains native to Europe, Africa, the Middle East and parts of central Asia. The term European honey bee refers to all *A. mellifera* subspecies or strains native to Europe. Various studies put the origin of *A. mellifera* either in Africa or the Middle East. Initially, however, it is likely that *A. mellifera* originated from *A. cerana*, although fossil and DNA records are currently insufficient to draw a firm conclusion. As more *A. cerana* and *A. mellifera* bees between Europe and South East Asia (where seven of the eight species of honey bees live) are genetically sequenced, a better picture of the relationship between the two species, as well as with other species of the honey bee, will emerge.

This book discusses pests and pathogens of the European honey bee, *A. mellifera*, which has been moved to many parts of the world due to its enhanced honey production, passivity and ease of management. Any strains of *A. mellifera* native to Africa, due to their aggressivity, would be classified as unwanted pests if they were to be found in other parts of the world.

Apis mellifera found in Africa

General description

African strains of *A. mellifera* are very aggressive; for example, *A. mellifera adansonii* from West Africa or *A. mellifera scutellata* from central and southern East Africa (Figure 10.2). Although any of the more than ten

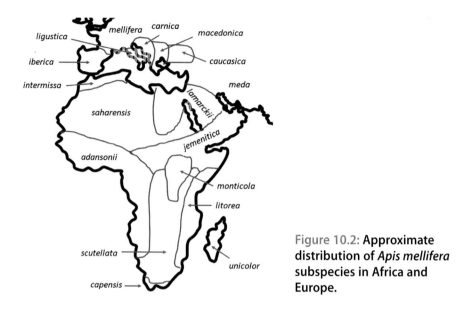

Figure 10.2: Approximate distribution of *Apis mellifera* subspecies in Africa and Europe.

strains of honey bees found in Africa can be called African honey bees, it is for *A. mellifera scutellata* that this name is usually applied.

Cape bee, *A. mellifera capensis*

The Cape bee or Cape honey bee, *A. mellifera capensis*, native to the Cape region of South Africa, would present a very different problem to beekeepers if it were to become established outside Africa (Figure 10.3). In South Africa, around the Cape, *A. mellifera capensis* beekeepers were successfully keeping this strain of honey bee for honey production. In 1990, to further exploit this strain's advantages, they took colonies of the Cape bee north to areas that were home to *A. mellifera scutellata*, commonly called the African honey bee. This created a unique problem because, unlike other strains of *A. mellifera* where workers can only lay unfertilized drone eggs, *A. mellifera capensis* workers can lay fully fertilized worker eggs that contain the worker bee's DNA, resulting in a viable worker being born without the *A. mellifera capensis* worker mating with a drone (a process referred to

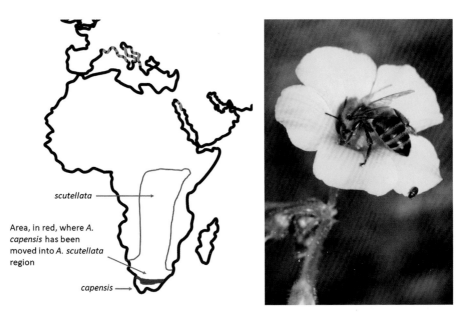

Figure 10.3: (a) Cape bee, *A. m. capensis* and *A. m. scutellata*, and where they share the same region. (b) Cape bee, *A. m. capensis* visiting a flower.

as automictic thelytoky).

When the Cape bee had been taken north into territory that was home to the African bee, Cape bee workers drifted into the hives of African bees, where they promptly started to lay viable workers: Cape bee brood. This led to a situation where inside an African bee colony there is both a queen African (*scutellata*) bee laying eggs and many Cape bee workers laying eggs. Soon, the number of Cape bees inside the colony overtakes the number of African bees, and the colony ultimately absconds or collapses. Alternatively, the colony may also produce a queen from a Cape bee worker-laid egg, which means that the colony has now changed from an *A. mellifera scutellata* colony to an *A. mellifera capensis* one. This situation is currently causing significant disruption to beekeeping in the northern parts of South Africa, and there does not appear to be a solution to this problem.

Not all the pathogens found infecting or infesting *A. mellifera* in the developed economies are prevalent across Africa. In December 2008

American foulbrood disease, *Paenibacillus larvae*, spread to the Cape bee population, infecting and wiping out an estimated 40 per cent of the region's honey bee population by 2015.

Africanized bees

The African honey bee, *A. mellifera scutellata*, was introduced to Brazil in 1956 and instrumentally mated with Italian bees, *A. mellifera ligustica*, imported to Brazil previously. Mated hybrids were kept in an experimental apiary of the University of São Paolo and 26 colonies were accidentally released when the queen excluder was removed from the entrance and the colonies swarmed. The escaped colonies slowly made their way north, eventually reaching the southern United States in 1990.

By the time hybrids had reached Texas, they had extensively interbred with previously introduced local strains of *A. mellifera* (such as Carniolan, Italian and Caucasian) and had become ferocious compared with the existing European bees. The cross of *A. mellifera scutellata* with the current varieties of South and North American *A. mellifera* has since been named Africanized bees, or killer bees, to distinguish them from the purer strains of African bees. If *A. mellifera scutellata*, *A. mellifera adansonii* or most other strains of African honey bees were to enter an area outside of Africa, the result would likely be very aggressive bees that would significantly threaten public health. Even today in West Africa, *A. mellifera adansonii* is so fierce that beekeepers work these bees only at night and change into their protective clothing before getting out of their vehicles.

Asian honey bee, *Apis cerana*

Apis cerana, commonly called the Asian honey bee, is widely used in Asia to produce honey. Confusingly, several other species of honey bees also have their home in Asia, and to call all of them Asian honey bees would be confusing (Figure 10.4). In this chapter we follow convention and call *A. cerana* the Asian honey bee, while all other honey bees found in Asia have been called either by their scientific name or by the collective name honey bees from (or in) Asia.

A. cerana is the most widely distributed honey bee in Asia and is extensively used in Asian countries to produce honey, wax and brood comb, which is eaten raw as a food source. *A. cerana* is called by many names in different parts of Asia, including the Asian honey bee, Asiatic bee, Asian hive bee, Indian honey bee, Indian bee, Chinese bee, eastern honey bee, Mee bee, and fly bee (Figure 10.5).

Across Asia, *A. cerana* can be found at higher altitudes, at sea level, in tropical, subtropical and temperate regions, in deserts, and in areas with high seasonal rainfall. The main area *A. cerana* has not been found is on a few isolated islands. As a result of this widespread distribution, different subspecies of *cerana* have very different temperaments and appearances.

Figure 10.4: **Size comparison of *A. cerana, mellifera* and *dorsata*.**

5 mm

In May 2007 the Asian honey bee was discovered in Cairns, Queensland, Australia. A colony was nesting in the mast of a boat that had recently arrived from Papua New Guinea. Subsequent searching revealed other colonies in the port area. By the time government apiary inspectors became aware of the incursion, there were multiple colonies of *A. cerana* located in and around Cairns. A concerted effort to eradicate the Asian honey bee was made, but the bee was declared endemic to Australia in 2011 since it was too difficult to eradicate.

Although *A. cerana* has its home across Asia, the particular race of *A. cerana* found in Cairns is *A. cerana javana*, which is native to Java (Indonesia), but can now be found in Papua (Indonesia), Papua New Guinea and the Solomon Islands.

Apis cerana is not as suitable as *A. mellifera* for large-scale beekeeping since they store less honey and have a higher inclination to abscond.

Asian honey bees moved into the Solomon Islands in 2003 and quickly destroyed the local *A. mellifera* honey industry. When the same subspecies of *A. cerana* was found in Cairns, there was concern that the bee would also destroy the Australian beekeeping industry. The Asian honey bee found in Cairns does not appear to be affecting the local European honey bee colonies except that they are competing for the same floral resources.

To date, most honey bee research has been conducted on the European honey bee subspecies of *A. mellifera*, and much less research has been conducted on the other seven species of honey bee. This is true for the Asian honey bee and particularly for the subspecies of *A. cerana* found in Cairns, *A. cerana javana*. The probable reason why this subspecies has not been studied is that it does not produce and store much surplus honey and is of very little economic importance in the areas of Java, New Guinea and the Solomon Islands where it can be found.

Figure 10.5: **Asian honey bee,** *Apis cerana.*

Characteristic	A. cerana	A. mellifera
Size	About two-thirds the size of A. mellifera	
Foraging range	Less than 2 km (1.25 miles)	Up to 5 km (3 miles)
Size of colony	Approx. 2600 in Cairns	50,000 to 60,000
Ability to pollinate	Yes	Yes
Aggressivity	Less aggressive than A. mellifera	
Resistance to Varroa, Tropilaelaps and tracheal mite	A. cerana can co-exist with Varroa without any impact	Significant health impact on A. mellifera
Swarming	Approx. three to six times a year	Once a year
Stripes on body	Distinctive stripes on abdomen	Not such distinctive stripes as A. cerana
Main differentiating characteristic	Vein pattern on rear wings	
Position on combs	A. cerana faces upwards when in a cluster and does the same when working comb.	A. mellifera faces downwards.

In Australia, the threat that *A. cerana* drones would mate with *A. mellifera* queens, forming a less useful hybrid, has not materialized. Studies have shown that 33 per cent of *A. mellifera* queens around Cairns have mated at least once with an *A. cerana* drone, while in parts of China, 14 per cent of *A. mellifera* queens have mated at least once with an *A. cerana* drone. Also, the Asian honey bee is moving out of Cairns at a rate of approximately 10 km (6 miles) per year. Overall, though, the long-term impact of this species of honey bee in Australia is not understood, due in part to significant inbreeding, resulting from a single founding mother at the time of incursion. As a result, the likely impact of the incursion is only a best guess from our limited knowledge of other subspecies of *A. cerana* in other parts of Asia.

Morphologically, *A. mellifera* and *A. cerana* are very similar in appearance and only an experienced observer is able to tell the difference between bees of the two species.

Although the life cycle of *A. cerana* is almost identical to *A. mellifera*, a worker emerges from capped cells after nineteen days for *A. cerana* and 21 days for *A. mellifera*. This small difference makes *A. cerana* much more resistant to *Varroa* compared to *A. mellifera*. In addition, *A. cerana* capped drone cells have a small pinhole in them but *A. mellifera* drone cells do not.

A. cerana are said to swarm more easily than *A. mellifera* and prefer to nest nearer to human habitation, such as in buildings or around cleared areas. Colony sizes are also much smaller, and in Cairns, the average colony size has been found to be about 2700 workers, compared with 50,000 to 60,000 for commercial *A. mellifera* colonies; the foraging range of *A. cerana* is about 2 km (1.25 miles), compared with 5 km (3 miles) for *A. mellifera*.

Like *A. mellifera*, *A. cerana* makes its nest in closed cavities like tree hollows and caves, or between rocks or in the walls of houses. The tropical strains of both *A. mellifera* and *A. cerana* make less honey than non-tropical strains and are also more prone to swarming than their subtropical cousins. This difference is due to several factors:

- **In the tropics there are more year-long supplies of nectar and pollen so that local bees do not need to store honey for times of scarcity.**
- **If there is a local shortage of nectar-producing flora near the nest, the colony can quickly abscond to a more suitable region, hopefully not too far away.**
- **There is often a greater threat from honey bee predators in the tropics and bees may abscond quickly as a defensive mechanism.**

A. cerana has far better grooming behaviour than *A. mellifera*, making it

Figure 10.6: Other honey bees from Asia. (a) *A. laboriosa*. (b) *A. dorsata*. (c) *A. dorsata* nest. (d) *A. florea*. (e) *A. florea* nest.

better at brushing off *Varroa* and other parasitic mites. *A. mellifera* does not possess good grooming skills but has superior hygienic behaviour, which is clearly illustrated in the way it removes infested brood from capped or uncapped cells and disposes of the infested larvae outside of the nest. Due to *A. cerana* and *Varroa* co-existing for possibly hundreds of years, the two species can live in the same colony with no known harmful effects on the bees. Over hundreds of years, *A. cerana* has developed effective defence mechanisms to manage infestations, while *A. mellifera* has not developed the same range of defence mechanisms and is thus more susceptible to *Varroa* infestations.

Across Asia, *A. cerana* is kept in hives by beekeepers to produce honey both to consume within their home and to sell for income. *A. cerana* is often preferred over *A. mellifera* by poorer people. This is because the set-up and ongoing management costs for *A. mellifera* are significantly higher than for *A. cerana*. If a European honey bee colony were to die or abscond, replacing the colony would be much more expensive than an *A. cerana* colony. A replacement *A. cerana* colony, if the original colony were to die or abscond, would be straightforward and probably almost free. The bee-keeper needs only to pay local children in their village a small amount of money to find a feral replacement colony nearby.

Figure 10.7: (a) Asian hornet, *Vespa velutina nigrithorax*. (b) Asian hornet nest. (c) Inside Asian hornet nest, with a larva that has been removed from its cell.

Other species of honey bees found in Asia

In addition to the species discussed above, some other common species are shown in Figure 10.6.

Asian hornets

Although not related to honey bees, the Asian hornet (*Vespa velutina nigrithorax*) is an invasive species in Europe, while the Asian giant hornet (*Vespa mandarinia*) is an invasive species in north-western North America.

Vespa velutina nigrithorax

The Asian hornet, *Vespa velutina nigrithorax*, although not found in North America, Australia or New Zealand, is causing significant problems to bee-keepers in France and other parts of Europe since its accidental introduction to that country, perhaps from China in 2004. Asian hornets are easily recognized since their thorax is velvety black or dark brown with distinctive yellow-orange stripes across it (Figure 10.7). Also, the head of the hornet is black with a yellow-orange face. The Asian hornet is larger than a European honey bee, with queens up to 30 mm in length and workers up to 25 mm long.

The Asian hornet makes its nest in the open, often hanging from trees, using a papier-mâché material made from chewed wood or plant material. The almost circular nest of the Asian hornet is easy to recognize as it is often about 1 metre (3 feet) in diameter.

The Asian hornet is a very aggressive predator of the European honey bee. Asian hornets will position themselves about 30 cm (1 foot) from the entrance to an *A. mellifera* hive and attack returning honey bees carrying pollen. The hornet will then take the honey bee to a nearby tree, where it will remove the legs and wings and turn the remaining parts of the honey

Figure 10.8: (a) Asian giant hornet, *V. mandarinia*. (b) Size of Asian giant hornet in comparison with human hand.

bee into a small meatball that they take back to their own nest as food for their larvae.

If the attack is severe, with multiple hornets attacking simultaneously, the effect on an *A. mellifera* colony can be catastrophic. The attack could, over a short period, lead to the destruction of the colony, or, in lesser attacks, the *A. mellifera* colony could be so weakened as to become vulnerable to other diseases, *Varroa*, or small hive beetle infestations.

Figure 10.9: **Asian and Asian giant hornets can be trapped by placing an attractant in a plastic bottle that has an open entrance, say cut into the side, that allows a hornet to enter but means it is unable to escape. Fresh meat can be used as an attractant as well as one of the commercially available liquid attractants. The bottle needs to be cleaned once a week and fresh meat or attractant added.**

Vespa mandarinia

The Asian giant hornet, *Vespa mandarinia*, is the world's largest hornet. It is native to mainland South East Asia and parts of the Russian Far East. The hornet has been found in Washington state, British Columbia and Vancouver Island. It was first found in Blaine, Washington, in 2019, with a few more additional sightings in 2020 and nests in 2021 (Figure 10.8).

Vespa mandarinia typically build underground nests, often in pre-existing cavities such as abandoned burrows or hollowed-out areas around the roots of trees. Above-ground nests are sometimes constructed in hollow tree trunks or similar cavities, though these nests are not usually more than a few metres from the ground.

The discovery of the hornet in North America is a concern since it preys on honey bees, wasps and other insects. *Vespa mandarinia* workers have an extensive foraging range and can attack prey located within 2 km (1.2 miles) of the nest. However, they have been found to travel as far as 8 km (5 miles) in search of food. If *V. mandarinia* were to become widespread across North America, the effect on the beekeeping industry, as well as hornet attacks on people and animals, would become a serious concern.

The hornet utilizes a pair of attack strategies to hunt bees and other insects. One strategy is for a single hornet to capture one honey bee at a time outside the entrance of a beehive. The hornet kills the bee by detaching the head from the rest of the body at the thorax, chews the thorax into a paste and takes it back to its colony to feed larvae. Hornets have a thick cuticle, or outer layer, which is mainly impenetrable by a bee's stinger. European honey bees are practically defenceless against this hornet since they have not evolved strategies to defend their colonies. However, Asian honey bees, *Apis cerana*, have developed defensive strategies due to their longer co-existence with the hornet. One technique involves many bees 'balling' a hornet to overheat and kill it, thus managing an attack.

Epidemiology

Introduction

Epidemiology studies how pathogens and diseases are distributed and spread through a population. The tools of epidemiology can be applied to humans and other animals, including bees, and provides an important tool to manage diseases on a large scale (region, country or continent) and inform national policy development, which impacts honey bee health worldwide.

Honey bees, *A. mellifera*, arguably live in a more complex, unfriendly environment than most other animals, in their wide geographic distribution, at the local level, and within hives. Managed colonies, particularly if they are frequently moved and/or if the beekeeper is inexperienced or sloppy, are often less healthy than feral colonies. As an example, American foulbrood is more common in managed than feral colonies since the use of contaminated tools by beekeepers, and the close proximity of hives within an apiary, facilitate transmission. Also, managed colonies kept by commercial beekeepers in the United States, Australia and Europe, and to a lesser extent elsewhere, are moved frequently to satisfy pollination contracts or to find new sources of nectar and pollen. This has the effect of 'seeding' infection into honey bee populations in new, previously disease-free areas.

Another difference between bees and other agricultural animals is that one of the main disease-control techniques used routinely in the management of infectious disease outbreaks in animals — the restriction of intentional movement of animals from one location to another — cannot be used. Indeed, foraging workers sometimes come in contact with other bees from many kilometres away and sometimes drift between hives. Hence, while disease-control authorities can place restrictions on the movement of hives in the event of an outbreak (which reduces the likelihood of long-distance disease spread) there are limited options available for preventing

Figure 11.1: **A colony of bees clustering at top of hive.**

short-distance ('local') spread. Hence, if 'detection and destruction' is used as a control strategy for eradication, it is essential to define a large area (say, 10 km/6 miles) around an affected beehive and destroy both managed and feral hives within that area. This fundamental characteristic of bees places a special responsibility on beekeepers. By taking steps to prevent the introduction of disease into the hives within your apiary, you're protecting the bees of your neighbours as well. This is particularly important in an urban setting where apiaries are typically within bee-flying distance of each other.

To study diseases of the honey bee, their causes and treatment, beekeepers and scientists need to understand the complex environment in which bees live, the pattern of disease within this environment, and the way the environment influences the distribution of pathogens and diseases. Without this knowledge, the management of pathogens and diseases is difficult, and here epidemiology can help understand and manage factors that influence honey bee population health. Epidemiological techniques used for honey bees can:

1. quantify the frequency of diseases at either the hive or apiary level
2. identify disease risk factors at the hive, apiary, beekeeper and/or geographical levels
3. develop a better understanding of the causes of disease based on frequency and risk factors, which in turn leads to the ability to
4. develop strategies to control or eliminate the disease from the population.

Epidemiological methods

Descriptive studies to determine the frequency of disease in bee populations

Quantifying the frequency of disease allows objective comparisons of the disease burden at the population level and across different population groups (such as commercial versus non-commercial apiaries, different geographic areas, managed versus feral colonies) and hence prioritized control efforts. Ongoing estimation of disease frequency over time allows changes in disease patterns to be identified; for example, the spread of disease into new areas or the emergence of new diseases or syndromes, such as Colony Collapse Disorder.

Two items of information are required to estimate the frequency of disease. First, the size of the population needs to be estimated. Ideally, this will be the number of hives, but sometimes it may be expressed as the number of apiaries. Second, the number of hives or apiaries in our population having the disease needs to be determined. Detection of the presence of disease might require a visit from an apiary inspector or a veterinarian, although beekeepers are often asked to submit the results of testing to authorities.

'Representativeness' is an important issue here, ideally requiring a full list of hives/apiaries in an area of interest to guide the testing regime and obtain a representative sample of hives and apiaries in the population. Focusing on specific areas or on specific subgroups of the population might lead to either an over- or underestimate in the frequency of disease, which may then lead to incorrect decision-making.

Figure 11.2 provides details of investigations carried out to determine the frequency of *Varroa destructor* among apiaries in the greater Auckland (New Zealand) area following the first detection of mites in April 2001. Figure 11.2a shows the location of all apiaries in the greater Auckland area as of April 2000. Figure 11.2b shows the location of apiaries that were tested for *Varroa* using the sticky board test, during April and May 2000. The results of sticky board testing are shown in Figure 11.2c. The apiary level prevalence of *Varroa*, based on sampling carried out in April and May 2001, was 28 per 100 apiaries at risk. Knowledge of apiary location, as well as the results of testing, allows high-risk disease areas to be distinguished from low-risk disease areas. To make it easier to interpret the data, Figure 11.2d is a heat map showing the number of *Varroa*-positive apiaries per 100 apiaries per square kilometre. In Figure 11.2d there's a relatively high density of *Varroa*-positive apiaries close to Auckland International Airport, providing an important clue regarding how *Varroa* entered the country.

There are two advantages to the heat mapping approach. First, the heat map provides a detailed representation of the geographic distribution of disease risk, sufficient to support decision-making (much more so than if we showed the prevalence of the disease by small area unit; for example, postcodes). Second, heat mapping eliminates concerns about privacy since we're not showing individual apiary locations on the map.

Some countries require beekeepers to report when hives are moved from one location to another. If a new disease is discovered, authorities need to determine where hives located near the outbreak have been taken to, or brought from, so they can quickly test those (source and destination) apiaries for signs of disease. In infectious disease outbreak management, this is called 'backward' and 'forward' tracing.

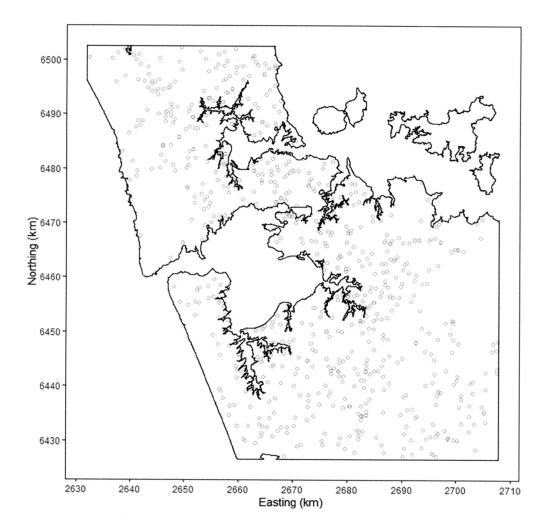

Figure 11.2: *Varroa destructor* in the greater Auckland area of New Zealand, April 2000: (a) Map showing the point location of all honey bee apiaries in the greater Auckland area in April 2000.

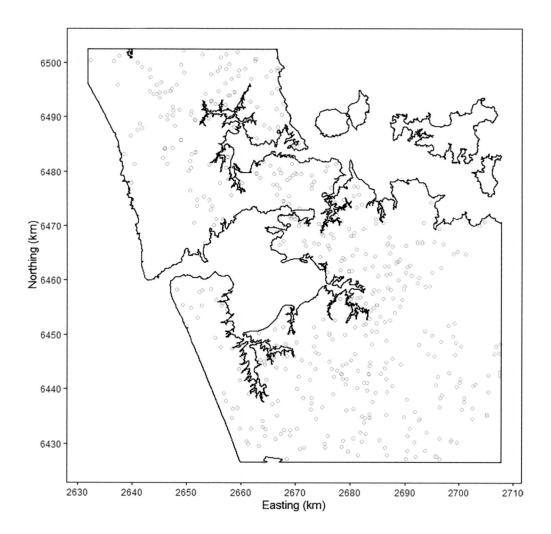

(b) Map showing the point location of apiaries tested for *Varroa destructor* in April and May 2000 using the sticky board test.

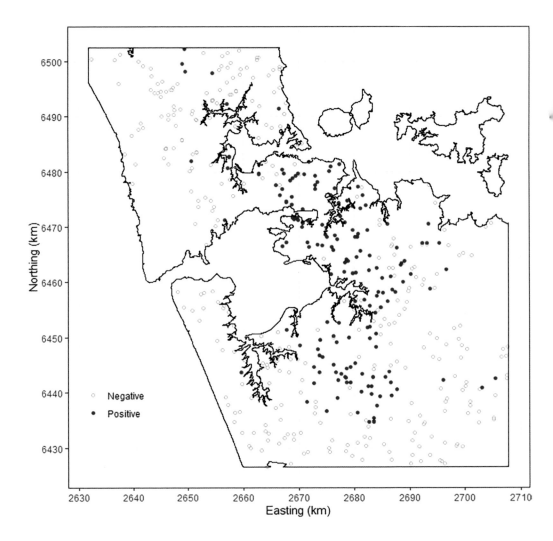

(c) Map showing the location of *Varroa* negative and positive apiaries, as determined by sticky board testing. The locations of individual apiaries have been jittered for privacy reasons.

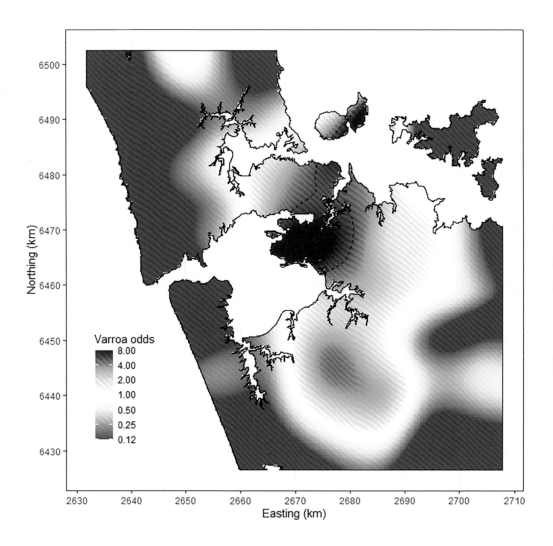

(d) Heat map showing the spatial distribution of *Varroa* odds (risk).
Areas marked in red are high risk; areas marked in blue are low risk.
Adapted from Stevenson et al. (2005).

The requirement to declare hive movements can present problems because beekeepers may not wish to declare commercially sensitive information such as the location of good nectar sources. Our advice here is simply to trust your animal health authority. Indeed, when registering as a beekeeper, authorities typically declare that the details shared with them (hive numbers, hive locations and details of movement) will be used only for emergency disease outbreak management. Accurate, precise and up-to-date location information enhances the effectiveness of outbreak response, which means that the negative impacts of disease on the industry as a whole will be minimized. Many beekeepers will not rent their colonies for the commercial pollination of crops since there is an increased risk of disease from other, nearby, commercial colonies.

Risk-factor studies and developing appropriate disease-control measures

Descriptive studies documenting the frequency of disease and how the frequency of disease varies according to the hive/apiary, place and time are useful for generating hypotheses ('theories') about why disease might be more common in some groups of a population compared to others. Once these theories have been developed, they can be tested by quantifying the association between a hypothesized risk and the prevalence of a given disease.

For example, in the case of American foulbrood, co-locating hives during pollination is known to spread the disease from one apiary to another. Some members of the beekeeping community might argue that disinfection of equipment is the best way to reduce the risk of disease while at the same time allowing the co-location of apiaries to take place. Given this information, we might carry out a study whereby we visit beekeepers, determine the American foulbrood status of their apiary, and ask them:

- **Do they co-locate their apiaries with other beekeepers?**
- **If they do, do they minimize the mixing of colonies by placing their apiaries in more distant parts of the crop?**

- **Do they disinfect all equipment and gloves prior to reuse?**

With this information, we calculate the prevalence of American foulbrood among apiaries where beekeepers:

- **Never co-locate apiaries or**
- **Co-locate apiaries and disinfect equipment prior to reuse, or**
- **Co-locate apiaries without disinfecting equipment prior to use.**

By comparing prevalence estimates among the three groups, we might find that American foulbrood among co-located apiaries, where all equipment was disinfected prior, was 1.5 times the prevalence of American foulbrood among apiaries that were never co-located. Similarly, the prevalence of American foulbrood among apiaries that were co-located and in which equipment was disinfected might be 4.5 times the prevalence among apiaries that were never co-located. This type of investigation (a cross-sectional study) is useful for quantifying the relative importance of 'exposures' (borrowing equipment in this example) on the relative risk of disease and allows recommendations for disease control to be fine-tuned. For the American foulbrood example, the key finding was that there was a substantial increase in disease risk if the equipment was shared without disinfection. An appropriate policy might be to not ban equipment sharing but to strongly encourage beekeepers to disinfect equipment prior to use, using, for example, an appropriately targeted disease-awareness campaign.

Often there are several factors that increase the risk of disease. Cross-sectional studies of the type described here are useful for working out which risk factor has the greatest impact on the level of disease in the population. The challenge for population health professionals is to determine the appropriate mix of achievable and acceptable management actions to reduce the risk of disease.

Geography

In earlier days, when most lands were used for diverse agriculture, there was a wide range of flowering plants on which bees could live. This land included uncleared areas with native vegetation and offered many sites for feral colonies. Bees were able to obtain a wide variety of foods, such as pollens, nectar, lipids, fats and minerals. The resulting diverse food supply enabled them to support a healthy life year-round. Migratory beekeeping was minimal, although earlier beekeepers relied on skeps (a straw or wicker beehive), clay pots and other forms of housing, that necessitated killing, or severely disturbing the colony, to remove the honey. Today, there is often far less diversity in flowering plants for managed colonies. This results in poor nutrition, which causes brood development problems, as well as nutritional deficiencies for adult bees. In extreme cases, if a colony is unable to collect sufficient food, adult bees will cannibalize larvae to provide food for the colony.

Migratory beekeeping is practised in many countries. This presents challenges because migratory beekeeping is the perfect method for spreading disease across a wide geographical area over a short period of time. When moving colonies from one type of monoculture to another, neither of which provide adequate nutrition for bees, diet may be a factor influencing the occurrence of disease. Alternatively, the stress of moving a colony frequently may be the most significant risk factor. As an example, studies in the United States show that hobby beekeepers have higher levels of *Varroa* than commercial beekeepers. However, hobby beekeepers have lower levels of nosemosis than commercial beekeepers.

The global spread of pathogens and diseases, usually caused by human action, shows that some bee populations are less susceptible to disease than other populations. Whether this is due to differences in climate, geography, food availability or genetics has not been clearly determined. The role of epidemiology is to tease out these factors and to better understand and manage the spread of diseases in a given population.

Summary

Epidemiology is used to help formulate hypotheses about disease occurrence and disease spread that can be tested experimentally; for example, by co-locating bees from different genetic backgrounds in each area and measuring disease susceptibility. By isolating one factor at a time or a small number of factors individually in an experimental set-up, it is possible to supplement epidemiological studies to further pinpoint and unravel the precise factors that drive diseases in certain populations. This untangling is often complex and time-consuming because many factors can play a role in some but not all situations. However, it is important to understand these factors as they may be critical for disease control. One example is the finding that the *Varroa* mite is not only directly detrimental to hive health but also is a vector for certain viral diseases, such as the deformed wing virus (DWV). The combination of *Varroa* mite with DWV is extremely lethal to bees. Hence, when DWV is present when *Varroa* is detected, as is almost always the case in *Varroa*-endemic areas, even relatively small levels of *Varroa* infestations are lethal to the hive. In contrast, when the viral load for DWV is low, or DWV is absent, the hive can tolerate much higher mite counts. This has obviously important implications for the management of the mites in different regions and at different stages of the *Varroa* infestation within these regions.

Hence, epidemiological studies are an important tool to identify colonies and/or apiaries that are more likely to have the disease and to guide actions during a disease outbreak that minimize the impact of the disease and maximize bee health, while at the same time allowing the industry to perform its role in terms of honey production and pollination services. Over time, for example, bee populations more resistant to disease may be selected, or treatments that were previously effective become obsolete due to pathogens developing resistance. Epidemiological studies will identify these changes and the priorities of the bee health authorities are likely to evolve in response. Therefore, epidemiologists play a continuing role in preserving honey bee population health.

Appendix A: Diagnostic table

	Bacteria			Fungi	
Pathogen	American foulbrood	European foulbrood	*Spiroplasma*	Chalkbrood	Stonebrood
	Paenibacillus larvae	*Melissococcus plutonius*	*Spiroplasma apis,* *Spiroplasma melliferum*	*Ascosphaera apis*	*Aspergillus flavus*
Appearance of Brood Comb	Sealed brood cells contain dead larvae/pupae. Cappings are sunken, greasy and punctured. May have an irregular brood pattern.	Dead brood larval cells are rarely capped. May affect sealed brood in heavy infections. May have an irregular brood pattern.	No symptoms.	May affect sealed or unsealed brood. Cell cappings may be punctured and contain mummies.	May affect sealed or unsealed brood. Hard yellow or greenish mummies.
Age of Affected Brood/Bee	Usually affects older, pre-pupae or pupae sealed brood.	Affects young, unsealed larvae that lie in a 'C' shape.	Adult bees.	Usually affects older larvae. Larvae are positioned upright in cells.	Usually affects larvae or pupae, and may infect adult bees. Larvae and pupae die in normal positions.
Dead Brood Shape and Colour	The larva or pupae within the capped cell will decay in place. Pupae are light brown, coffee-coloured to dark brown.	Progression of colour: from white to yellowish/ white then brown to dark brown/ black.	No symptoms.	Dead larvae have a hard, chalk-like appearance. Mostly white, although could be black or grey.	The body may be yellow to grey-green.

Viruses					Mites
Nosemosis	Sacbrood	Black queen cell virus	Deformed wing virus	Chronic paralysis virus, acute bee paralysis virus, Kashmir bee virus, Israeli acute paralysis virus	Tracheal mite
Nosema apis, Nosema ceranae	Sacbrood virus	BQCV	DWV	CPV, ABPV, KBV, IAPV	*Acarapis woodi*
No visible signs. *Nosema* is often spread by faeces infecting comb.	Cappings of sealed brood are punctured.	No symptoms.	No symptoms.	No symptoms.	No symptoms.
Adult bees.	Usually affects older larvae. Larvae are positioned upright in cells.	Queen pre-pupae or pupae may show symptoms. Can affect larva, pupa, and adult bees without showing symptoms.	Pupae and adults may be infected. Not all infected adults show symptoms of deformed bodies.	Pupae and adults may be infected. Not all infected adults show symptoms.	Adult bees.
No symptoms.	Larvae die on their backs with their head slightly elevated. Colour may progress from white to yellow then brown to dark brown.	The wall of the queen cell becomes dark brown to black after the death of the queen brood.	No symptoms.	No symptoms.	No symptoms.

Honey Bee Pests and Diseases

	Bacteria			Fungi	
Consistency of Dead Brood	Soft and becomes sticky and stringy	The consistency of the dead brood is watery (rarely sticky or stringy) and granular.	No symptoms.	Paste-like, before becoming hard pellets.	The body may be hard and difficult to crush.
Odour of Dead Brood	'Fish-like' odour. Strong odour.	Variable from none to sour.	None.	No odour to slightly non-objectionable.	None.
Larva/ Scale/Bee Appearance	The tongue of dead larva is often left attached to the top of the cell. Decayed brood form a dark brown or black brittle scale that adheres tightly to cell wall. Pupal tongue may be evident.	Dead larvae remain soft, pliable and usually twist/ curl in the cell. Workers can easily remove the cadavers. Not as hard or dark as AFB scale.	Neurological problems displayed by infected bees.	Mummies are loose and do not adhere to cell wall. Mummies are hard and smooth.	Fungus grows quickly and rapidly, turning the infected larva or pupa white. Eventually the cadaver turns brownish or yellow-green and becomes hard. After death, the body hardens. Fungus eventually forms a false skin, covering the cadaver with powdery fungal spores.
Identification	Matchstick test on decomposed pupal strings out about 3 cm (1 inch).	Minimal ropiness with matchstick test. Young larvae twisted and distorted in cell with visible tracheae. Darkened larvae.	Heavy die-off of adult bees, often away from the colony. Foragers are disoriented. Shaking of adult bees.	Most mummies are white or mottled. Some mummies may be black. Hard mummies scattered throughout brood cells, on floor of hive and hive entrance.	Symptoms are similar to chalkbrood. disease of the larvae and occasionally pupae.

	Viruses					Mites
No symptoms.	Consistency of dead brood is watery and granular inside the tough leathery sack.	Tough, sac-like skin forms around pupae.	None.	None.	None.	No symptoms.
None.	No odour to slightly sour.	None.	None.	None.	None.	None.
No symptoms.	Cadavers if not removed by the workers deflate uniformly, dry into a dark scale. Unlike American foulbrood, the scale does not adhere to the cell and can easily be removed. Consistency of dead brood is watery and granular inside the tough leathery sac.	Infected pupae die, darken, and the walls of the cell develop black patches.	Stubby, useless wings. Shortened, rounded abdomens, colouring. Paralysis	Adults may tremble, become bloated with shiny skin and with little hair. Unable to fly and may have disjointed wings. Pupal and adult may display paralysis.	No symptoms. Infests adult bees.	
Poor population build-up. Low honey production. Supersedure of queen in highly infected colonies. Possible diarrhoea on front of hive. Crawling bees (*N. apis* only).	Consistency of dead brood is watery and granular inside the tough leathery sac.	Dead queen pupae with black cell walls. BQCV needs co-infection with *Nosema* to spread BQCV to host bee.	DWV tends to remain in low levels in healthy colonies and exists as a low-grade infection with no symptoms. Adult bees may have shrivelled wings and bodies. Often spread by *Varroa*.	Adult bees tremble, and crawl around. ABPV, KBV and IAPV are often associated with *Varroa* infestations.	Many bees at the entrance are unable to fly and crawl around the entrance.	

Appendix B:
Varroa control options by seasonal phase

Different *Varroa* control options are appropriate for each of the four population phases of the honey bee/*Varroa* mite seasonal cycle. Below is a summary of options for each seasonal phase.

DORMANT PHASE

Bees are clustered; no brood in northern locations with reduced brood rearing in southern locations; all or most *Varroa* mites are phoretic (i.e. on adult worker bodies, as there is little to no developing brood) and both populations are in decline because there is little or no reproduction occurring within the colony.

Highly Effective Options:	Notes:
• Oxalic acid (fumigation method) • Winter or broodless period	• Oxalic acid is best used when there is no brood. • *Varroa* mortality over extended broodless period is high.
Moderately Effective Options:	**Notes:**
• HopGuard® II • In beekeeping regions with brood during this phase, Apiguard or Api Life Var® or formic acid (MAQS®), provided temperatures are within optimal ranges.	• Little or no independent test results are available for HopGuard® II during the dormant phase. The formulation has changed each of the last two years. • The effectiveness of Apiguard®, Api Life Var® and formic acid (MAQS®) during the dormant phase is unknown.
Least Effective Options:	**Notes:**
• Anything that risks colony success through this phase • Screen bottom board	• Screen bottom board removes a small percentage of mites that fall from adult bodies. It is best used in combination with other techniques.

POPULATION INCREASE

Seasonal colony build-up; colony brood population growing rapidly and adult worker population increasing; *Varroa* mite population usually low but increasing; pre-honey flow supering of colonies.

Highly Effective Options:	Notes:
• Apivar® • Apiguard® or Api Life Var® • MAQS® (formic acid) • Drone brood removal	• Terminate Apivar® after 42 to 56 days of treatment, at least two weeks prior to adding supers. • Terminate Apiguard® treatment before adding supers. • Terminate Api Life Var® after two or three treatments (7–10 days each). Remove Api Life Var® tablets from the hive at least one month before harvesting honey or, if not using the colony for honey production, treat for full treatment period. • It is legal to use MAQS® when storing honey. • Strong, populous colonies tolerate drone brood removal two to three times.
Moderately Effective Options:	Notes:
• HopGuard® II • Colony division • Requeening using hygienic stock • Basic sanitation	• The effectiveness of HopGuard® II has not been widely tested. • Dividing the colony during the population increase phase will most likely negatively affect surplus honey production. • Hygienic queens are not always available. • Basic sanitation may help reduce other stressors.
Least Effective Options:	Notes:
• Screen bottom board • Powdered sugar • Mineral oil • Failure to perform managements	• A screen bottom board is marginally effective. • There is little evidence that powdered sugar or mineral oil has any effect on mite populations.

POPULATION PEAK

Period of nectar flow and rental of colonies for pollination services; bee population (both adult and brood) at peak; mite populations increasing, nearing peak; often honey supers on colonies.

Highly Effective Options:	Notes:
• MAQS® • Apivar® or Apiguard® or Api Life Var® (Use is permitted only if no supers are present or colonies are not producing honey.)	• MAQS®, Apiguard® and Api Life Var® are not suitable for use in all temperatures. See the detailed descriptions of products below for temperature ranges for use of these products. • Apivar® (Amitraz) is highly effective. Be cautious about using it too often to avoid risk of developing resistance.
Moderately Effective Options:	**Notes:**
• Requeening with hygienic stock • Division of colonies • HopGuard® II • Oxalic acid drip	• Requeening or dividing may negatively affect honey production (if colonies are strong enough to produce surplus). Hygienic or locally selected stock is not widely available. • The effectiveness of HopGuard® II has not been widely tested. • Oxalic acid is best used when there is little or no capped brood in the colony during the dormant phase or because of queen replacement that interrupts brood rearing.
Least Effective Options:	**Notes:**
• Screen bottom board • Drone brood removal	• A screen bottom board removes a small percentage of mites that fall from adult bodies. Use it in combination with other techniques. • Drone brood removal is restricted in this phase by the absence of sufficient drone brood and the difficulty of accessing the brood nest beneath honey supers.

POPULATION DECREASE

Post-honey harvest; bee population decreasing; colonies rearing overwintering bees. *Varroa* mite populations growing, peaking and then declining until eventually only phoretic mites on adult bees after colonies become broodless.

Highly Effective Options:	Notes:
• Apivar® • MAQS® • Apiguard® or Api Life Var® • HopGuard® II	• Apivar® should not be used until surplus honey is removed. • MAQS®, Apiguard® and Api Life Var® are not suitable for use in all temperatures. See the detailed descriptions of products below for temperature ranges for use of these products. • HopGuard® II manufacturer's test data supports its effectiveness.
Moderately Effective Options:	Notes:
• Requeening with hygienic bees • Dividing colonies • Oxalic acid drip	• Hygienic stock is not widely available. • Requeening and dividing colonies may be difficult. • Oxalic acid is most effective if there is little to no capped brood present.
Least Effective Options:	Notes:
• Apistan® or CheckMite+® • Drone brood removal • Screen bottom board • Sanitation	• Mite resistance to Apistan® and CheckMite+® is well established. • Colonies are unlikely to raise drones during this phase. • Basic sanitation may help relieve stress.

From: *Tools for Varroa Management: A Guide to Effective Varroa Sampling & Control*. Honey Bee Health Coalition. https://honeybeehealthcoalition.org. Reproduced with permission.

Appendix C: Integrated Pest Management

CATEGORY	NAME	BIOLOGICAL
Viruses	All	
Bacteria	American Foul Brood	
	European foulbrood	
Fungi	*Nosema cerana*	
	Nosema apis	
	Chalkbrood	

CULTURAL	CHEMICAL
Ensure colony has adequate pollen and nectar and other diseases are kept to a minimum.	There are no approved chemical treatments for bee viruses.
Ensure colony has adequate pollen and nectar.	Oxytetracycline (OTC) is approved in some countries.
Quarantine hive.	
Destroy colony and burn or irradiate hive.	
Requeen infected and non-infected colonies with resistant stock.	
Minimise robbing of weak colonies by reducing size of entrance.	
Quarantine swarms before introducing them into the apiary.	
Replace two frames of dark brood comb every year with clean foundation.	
Hygienic apiary management practices.	
Ensure colony has adequate pollen and nectar.	Oxytetracycline (OTC) is approved in some countries.
Requeen infected and non-infected colonies with resistant stock.	
Hygienic apiary management practices.	
Ensure colony has adequate pollen and honey stores, particularly during the winter.	Fumagalin B is approved in some countries.
Leave hive in warm, dry location. Leave alone for several months with minimal monitoring.	
Requeen with resistant stock.	
Keep inside of hive dry.	
Locate hive in a sheltered position.	
Strengthen weak colonies.	
Minimize the amount of time that the colony are contained within the hive since they need regular flights to defecate outside of the hive.	
Night insect zappers.	
Replace two frames of dark brood comb every year with clean foundation.	
Requeen with resistant stock.	There are no approved chemical treatments for chalkbrood.

Honey Bee Pests and Diseases

CATEGORY	NAME	BIOLOGICAL
Parasitic Mites	*Varroa*	Powdered sugar dusting promotes bee grooming. (Could be classed as cultural).
	Tropilaelaps	Pseudoscorpions.
		Velvet mite.
		Entomopathogenic fungi.
		Parasitic nematodes.
	Tracheal mite	
Insect Pests	**Small hive beetle**	Carbid beetles.
		General soil predators that eat larvae.
		Spiders.
		Lizards.
		Parasitic nematodes.
	Greater wax moth	Parasitic wasps.
	Lesser wax moth	Birds.
		Bacteria.

CULTURAL	CHEMICAL
Keep inside of hive dry.	
Strengthen weak colonies.	
Locate hive in a sheltered position.	
Replace badly affected frames with new foundation.	
Requeen with resistant stock.	See Chapter 3 on *Varroa* for list of management techniques.
Minimize brood comb.	
Screened bottom boards.	Chemicals used to manage *Varroa* are expected to be effective against *Tropilaelaps*.
Only use *Varroa*-resistant bees, e.g.Varroa Sensitive Hygienic (VSH) or Russian Stock.	
Requeen with resistant stock.	See chapter on tracheal mite for list of chemicals that have been used for management.
Ensure colony has adequate pollen and honey stores, particularly during the winter.	
Minimize robbing of colonies by reducing size of entrance.	
Quarantine swarms before introducing them into the apiary.	
Maintain strong colonies.	See chapter on small hive beetle for list of chemicals that have been used for management.
Remove unused or over-applied pollen patties.	Fiprinol is frequently used for management.
Hygienic apiary management practices.	
Maintain clean honey house.	
Render all wax cappings and debris quickly after the extraction.	
Store supers in a secure location before they are returned to colonies.	
Freezing frames will kill small hives beetles.	
Store equipment in a cool storeroom below 4°C.	Aluminium arsenate may be used to sterilize infected empty hives. Check regulations in your countrty.
Night insect zappers.	
Remove comb not covered in bees.	
Use new comb.	
Keep hives strong.	

Bibliography

Pests and diseases

Abrol, D.P. (2008). *Honeybee Disease and their Management.* Kalyani Publishers.

Applegate, J.R. & Britteny, K. (2021). *Honey Bee Veterinary Medicine:* Apis mellifera. Elsevier. ISBN: 978-032-3896863

Aston, D. & Bucknall, S. (2021). *Good Nutrition, Good Bees.* Northern Bee Books. ISBN: 978-191-2271955

Aubert, M.F.A. (2011). *Virology and the Honey Bee.* Directorate-generale for Research, European Commission. ISBN: 978-927-9005862

Basterfield, D., Cullum-Kenyon, R. & Davis, I. (2019). *The Healthy Hive Guide.* Northern Bee Books. ISBN: 978-086-0982890

Bunker, S. (2019). *The Asian Hornet Handbook.* Sarah Bunker. ISBN: 978-191-6087101

Caron, D.M. (2018). *Tools for Varroa Management: A guide to effective Varroa sampling & control.* https://honeybeehealthcoalition.org/wp-content/uploads/2018/06/HBHC-Guide_Varroa_Interactive_7thEdition_June2018.pdf

Carr, J. (2016). *Managing Bee Health: A practical guide for beekeepers.* 5M Publishing. ISBN: 978-191-0455036

Goodwin, M. & Taylor, M. (2021). *Control of Varroa: A guide.* New Zealand Ministry of Agriculture and Forestry. ISBN: 978-0473-570774

Gregory, P. (2018). *Healthy Bees are Happy Bees: A comprehensive guide to bee health and sickness.* Bee Craft Limited. ISBN: 978-090-0147302

Hansen, H. & Morse, R. (1980). *Honey Bee Brood Diseases.* Wicwas Press. ISBN: 978-878-7905046

Heaf, D. (2021). *Treatment-Free Beekeeping.* IBRA & NBB. ISBN: 978-191-3811006

Hepburn, H.R. & Radloff, S.E. (2012). *Honeybees of Africa.* Springer. ISBN: 978-366-2036051

Hepburn, H.R. & Radloff, S.E. (2014). *Honeybees of Asia.* Springer. ISBN: 978-364-2422829

Hesbach, W. (2016). *Splits and Varroa: An introduction to splitting hives as part of Varroa control.* Northern Bee Books. ISBN: 978-190-8904867

Honey Bee Health Coalition. Website: https://honeybeehealthcoalition.org/

Hood, W.M. (2017). *The Small Hive Beetle: Aethina tumida Murray.* Northern Bee Books. ISBN: 978-191-2271078

Kane, T.R. & Faux, C.M. (2021). *Honey Bee Medicine for the Veterinary Practitioner.* Wiley-Blackwell. ISBN: 978-111-9583370

Lester, P. (2021). *Healthy Bee, Sick Bee: The influence of parasites, pathogens, predators and pesticides on honey bees.* ISBN: 978-177-6564057

Martin, S. (2017). *The Asian Hornet: Threats, biology & expansion.* IBRA. ISBN: 978-086-0982814

McMullan, J. (2021). *Having Healthy Honeybees.* Northern Bee Books. ISBN: 978-191-2271900

Morse, R. & Flottum, K. (2013). *Honey Bee Pests, Predators, and Diseases.* Northern Bee Books. ISBN: 978-191-4934070

Nagaraja, N. & RajaGopal, D. (2009). *Honey Bees: Diseases, parasites, pests, predators and their management.* MJP Publishers. ISBN: 978-818-0940590

Oldroyd, B. & Wongsiri, S. (2006). *Asian Honey Bees: Biology, conservation, and human interactions.* Harvard University Press. ISBN: 78-0674021945

Pernal, S. & Clay, H. (2013). *Honey Bee Diseases & Pests.* Canadian Association of Professional Apiculturists (CAPA)

Sharma, V. (2014). *Honey Bee Diseases & Their Management.* Random Publications. ISBN: 978-935-1113584

Somerville, D. & Annand, N. (2014). *Healthy Bees: A practical handbook — managing pests, diseases and other disorders of the honey bee.* NSW Department of Primary Industries, Australia. ISBN: 978-174-2566009

United States Department of Agriculture. *Diagnosis of Honey Bee Diseases.* USDA.

Vidal-Naquet, N. (2015). *Honeybee Veterinary Medicine:* Apis mellifera L. 5M Publishing. ISBN: 978-191-0455043

Epidemiology

Elwood, M. (2018). *Critical Appraisal of Epidemiological Studies and Clinical Trials.* Oxford University Press. ISBN: 978-019-9682898

Stevenson, M. et al. (2008). *Spatial Analysis in Epidemiology.* Oxford University Press. ISBN: 9780198509882

Thrusfield, M. & Christley, R. (2018). *Veterinary Epidemiology, 4th Edition.* Wiley-Blackwell. ISBN: 978-1-118-28028-7

Acknowledgements

A book such as this could not be written without the assistance of many people. Pests and pathogens of the honey bee is a wide area and few people have a grasp of the entire field. If the book is of use to beekeepers, thanks are due to the many contributors; if there are mistakes or omissions in the book, the authors take responsibility. Special thanks go to Ms Bron Woods, of Watsons Creek, Victoria, Australia, for reading the manuscript in its entirety and offering many useful suggestions. Dr John Roberts of the CSIRO, Australia, for reading the chapters on how pathogens infect bees, viruses and *Varroa*, three important chapters. Dr James Sainsbury, of Plant & Food Research, New Zealand, provided information on brood diseases, while Dr Anna Gajda, Institute of Veterinary Medicine, Warsaw, Poland, provided information for parts of the chapter on viruses. Dr Michael Bentley, in the United States, contributed his extensive knowledge of other types of bees and hornets to that chapter. Too many people have contributed their time and expertise to mention every person's role individually. I hope that they are not offended by their omission, but their contribution was equally appreciated by the authors.

Appendix B, Varroa Control Options. Reproduced with permission. Honey Bee Health Coalition. *Tools for Varroa Management: A guide to effective varroa sampling & control.* https://honeybeehealthcoalition.org

Photographs and illustrations

Bentley, Michael: 2.4, 2.8, 11.1

CSIRO, Australia: 2.10b

de Gelder, Klaas: 3.7

El Ministerio de Agricultura, Spain: 3.2

Encyclopedia of Life: 8.13c, 8.13d

FAO: 1.3

FERA, UK: 7.7b

Gajda, Anna: 7.3, 7.7a

Garvey, Kathy Keatley: 2.2

Huang, Zachary: 4.1c, 5.2b, 5.2c, 7.1a

Imhoff, Markus: 2.9

Kryger, Per: 5.7c

Manning, Rob: 3.6c

McAfee, Alison: 3.5

National Bee Unit, UK: 4.1b

Neumann, Peter: 8.5b, 8.5c, 8.5d

Nolan, Kate: 2.3, 2.6, 3.3, 3.4, 4.1a, 4.2,
    4.5, 5.5b, 6.1, 8.1

Oldroyd, Ben: 8.11a, 8.11b, 8.11c, 8.11d,
    10.6c, 10.6e

Ontario, Government of: 2.10a, 3.10a

Owen, Barbara: 2.7

Owen, Robert: 0.1, 0.2, 1.2, 2.1, 2.2, 2.5,
    5.3a, 5.3b, 5.4, 5.5, 5.6, 8.3a, 8.3b,
    8.3c, 8.4a, 8.4b, 8.8, 8.9, 8.10, 10.2,
    10.3b

Pernal, Steve: 7.2b

Ptaszyńska, Aneta: 6.2a, 6.2b, 6.2c

San Martin, Gilles: 3.1, 10.7c

Scheerlinck, Jean-Pierre: 3.8a, 3.8b,
    3.8c, 3.8d, 3.8e, 3.8f, 3.8g, 3.9a,
    3.9b, 3.9c, 3.9d

Stevenson, Mark: 11.2a, 11.2b, 11.2c,
    11.2d

Topolska, Grażyna: 7.1b, 7.1c, 7.2, 7.3b,
    7.3c, 7.5, 7.6

Trunch bees, UK: 3.11a, 3.11b

Underwood, Robyn: 3.10b, 3.10c

University of Florida: 6.3, 8.12

USDA: 4.3, 5.1a, 5.1b, 5.4a, 5.4b, 5.5a,
    6.4, 8.5a

Vita Europe: 8.7a, 8.7b

Walker, Ken: 10.4a, 10.4b

Wikipedia, Creative Commons: 10.5,
    4.1b, 4.1d, 5.2a, 7.5, 8.2a, 8.2b,
    10.6a, 10.6b, 10.6d, 10.7a, 10.7b,
    10.8a, 10.8b

Woods, Bron: 0.1, 1.1

Workman, Julie: 10.3a

www.beeaware.org.au: 3.6b

www.beehealth.uada.edu: 8.6a

www.beeinformed.org: 9.1a, 9.1b

www.beesource.com: 9.2

www.honeybeesuite.com: 9.3

Zhang, Miles: 8.13a

INDEX